W9-AXF-597

Praise for *The Social Life of DNA*

"Alondra Nelson tells a story for anyone interested in their own family, even their own memory. Using fresh genetics research and writing like an investigative reporter, Nelson clears up the mystery about our society's rush to DNA."

—Edward Ball, author of *Slaves in the Family*

"Alondra Nelson's account of how genetic data was transformed into contested political culture is as lucid as it is path-breaking. This exhilarating survey of how DNA became an agent in the politics of reparation and reconciliation has not only extended analysis of race and racism but created a new field of comparative research."

—Paul Gilroy, professor of American and English literature, King's College, London

"*The Social Life of DNA* is remarkable. Alondra Nelson explains the ways in which genomic research is being deployed by people, particularly black Americans, to explore their genealogical roots. She also examines the meaning of this activity: the motivations behind it, the communities it creates, the conflicts it engenders, the enterprises it supports, the changes in self-definition it encourages. Nelson explores this large, sprawling, fascinating subject with clarity, passion, rigor, and a keen eye for revealing detail. *The Social Life of DNA* will appeal to a broad readership interested in history, race, and science. Geneticists, sociologists, anthropologists, political scientists, and jurists will be stimulated by reading this book. It is a brilliant work."

—Randall Kennedy, Michael R. Klein Professor at Harvard Law School and author of *The Persistence of the Color Line*

"Alondra Nelson takes us into a complex and endlessly fascinating space where genetic ancestry testing meets racial politics. With her unique and wonderful gifts for research and insight into genetic science, ethnography, and history, *The Social Life of DNA* comes at a moment when the questions it raises about race and social justice couldn't be more pressing and urgent."

—Rebecca Skloot, author of *The Immortal Life of Henrietta Lacks*

"'The double helix now lies at the center of some of the most significant issues of our time,' Alondra Nelson writes in this valuable and illuminating book. Since 2003, she has been following the ways that DNA intertwines with race, and *The Social Life of DNA* is her clear-eyed, sharp, and closely observed account of the phenomenon. It couldn't be more timely."

—Jonathan Weiner, Maxwell M. Geffen Professor of Medical and Scientific Journalism at Columbia Journalism School

"One of this generation's most gifted scholars examines the unfolding mysteries of DNA sequencing and the limits and promises of genetic genealogy at the intersection of race, politics, and identity. Alondra Nelson brilliantly guides us on a journey of discovery in this cautionary tale of the high-stakes efforts to reconcile our racial origins and to find redemption as a country. Eye-opening, provocative, and deeply humane."

—Isabel Wilkerson, author of *The Warmth of Other Suns*

"*The Social Life of DNA* is a brilliant ethnography of the recreational uses of DNA. Besieged as our culture has become by beguiling promises of romantic heraldry and forensic infallibility, Nelson takes an unflinching yet sympathetic look at how popular yearning for 'lost roots' has led to DNA as metaphor: 'reading' our genes has become an inferential, often scientifically unsubstantiated link between past, present, and future. It has emerged as the symbolic grounding for magical cures, heritage tourism, and escapist fantasy, as well as legal actions for ethnic and racial reconciliation, reparations, and repatriation. Timely and original, this book offers a nuanced and engrossing negotiation between genetic truth and 'truthiness.'"

—Patricia J. Williams, James L. Dohr Professor of Law at Columbia University and columnist for *The Nation*

THE SOCIAL LIFE OF DNA

THE SOCIAL LIFE OF DNA

RACE, REPARATIONS, AND RECONCILIATION AFTER THE GENOME

Alondra Nelson

BEACON PRESS
BOSTON

Beacon Press
Boston, Massachusetts
www.beacon.org

Beacon Press books
are published under the auspices of
the Unitarian Universalist Association of Congregations.

Printed in the United States of America

19 18 17 16 8 7 6 5 4 3 2 1

This book is printed on acid-free paper that meets the uncoated paper
ANSI/NISO specifications for permanence as revised in 1992.

Text design by Wilsted & Taylor Publishing Services

Some names and other identifying characteristics of people mentioned
in this work have been changed to protect their identities.

Library of Congress Cataloging-in-Publication Data

Nelson, Alondra, author.
The social life of DNA : race, reparations, and reconciliation after the genome / Alondra Nelson.
pages cm
Includes bibliographical references and index.
ISBN 978-0-8070-3301-2 (hardcover : alk. paper) — ISBN 978-0-8070-3302-9 (ebook) 1. African Americans—Race identity. 2. African Americans—History. 3. Genetics—Social aspects—United States. 4. Genealogy—Social aspects—United States. 5. Genomics—Social aspects—United States. I. Title.
E185.625.N45 2016
305.896'073—dc23 2015025304

To my parents,
Robert Samuel Nelson and
Delores Yvonne Nelson,
who gave me roots and wings

CONTENTS

PREFACE

Like many Americans, my family and I were riveted by the *Roots* miniseries when it first aired in January 1977. I vividly recall sitting in front of the television with my mother, father, sister, and two brothers watching the story of Alex Haley's family unfold in Technicolor.

My father, having just completed a tour at sea, reclined in an armchair, his feet up. My mother was on the sofa with one or two of us kids twined tightly around her. The other two of us were on the floor, alternately being admonished by our parents not to lie too close to the screen or told, courtesy of a sibling, to move out of the way. On Sunday evening, when it became apparent that we would view the first episode in its entirety—well past our bedtimes—we knew we were in uncharted territory.

The *Roots* occasion provided one of those unforgettable moments when a child sees her parents in a new light. Watching *Roots*, I also watched my parents, who were visibly stirred by Haley's account. More than a few times during those eight evenings, my mother's eyes welled with tears. She frequently shook her head and murmured "*Uhm-uhm-uhm*," as I had heard Mary, her Philadelphia-born mother, do many times. An inherited response for emotions that defy language, perhaps. My father, who hailed from New Orleans, was characteristically stoic, but occasionally allowed a "That's a damn shame" during an especially graphic or tragic scene. I realize now that while watching *Roots*, my parents similarly watched us, their children. They were worried and protective, interspersing their own commentary between scenes, hoping to ameliorate the dramatic effect of this painful history.

The *Roots* effect expanded beyond our family home, perched on the edge of a craggy San Diego canyon, to my grade school, nestled in a valley. I was called Kizzy and Kunta Kinte by my mostly blond classmates during first period at my Southern California private school. But during our lunch breaks, the teasing gave way to earnest but clumsy conversations. In the schoolyard, we tried to make sense of what *Roots* meant for our interracial friendships, for our discussions in Sister Nora's American history class, and for our nation in the wake of its bicentennial. In our own ways, we each wondered, Who are we in relation to this history? Did this really happen? If so, how did we get from then to now—and where do we go from here?

Haley made his mark as a collaborator on *The Autobiography of Malcolm X*, the late activist's influential account of his political transformation published in 1965. This work emerged at the beginning of the black power era. *Roots*, published in 1976, and the television miniseries that was based on it, which premiered a year later, were culminating symbols of the era. This was the time of the Afro and the dashiki—of the "Black is beautiful" ethos. Between 1965 and 1977, black Americans turned to their African origins with intensity.

This interest in African origins and, in turn, genealogy was piqued in 1977. This watershed year also saw the publication of *Black Genealogy* by Charles L. Blockson, a primer of root-seeking attuned to the needs of African Americans, who faced especially steep hurdles in tracing ancestry. The Afro-American Historical and Genealogical Society (AAHGS), the first national black organization dedicated to genealogy and family history, was also established in 1977. In the intervening decades, genealogy only grew in appeal for African Americans. In the last decade, with the decoding of the human genome, new tools were introduced that expanded the popularity of genealogy exponentially and, moreover, gave it multifaceted uses.

I began research for this book in 2003 after noting mention in the press of a DNA testing service that promised to help blacks trace their roots. I was captivated. At that time, genetic ancestry testing was in its infancy and traditional gatherings of genealogists were where the early adopters of these new root-seeking techniques could be found. I attended scores of these gatherings, large and small, throughout the country from Oakland, California, to Bedford, Massachusetts, and numerous places in between. My travels also took me to the United Kingdom. In these places, I encountered genealogists who had been using archives and oral history to reconstruct their family stories and who were willing to try the new genetic-ancestry-testing services that were just hitting the market.

I've also participated in events and conferences at which genetic genealogy testing was discussed, including meetings at churches, libraries, and universities, and conducted fieldwork and interviews in settings both virtual and concrete. I interacted with genetic genealogists and eventually, in a now well-established tradition of social science research called "participant observation," I also became a root-seeker. I started conducting research on my own family's history, which besides Pennsylvania and Louisiana traverses parts of the southern United States as well as the country of Jamaica, and became a card-carrying member of the Afro-American Historical and Genealogical Society.

Building a bridge to Africa has inspired black American arts, letters, and politics for generations. Even if these speculative "roots" tests I read about never materialized, here a cutting-edge answer was being proposed to a central enigma of African America—a remedy that seemed ripped from the pages of a sci-fi novel. Speculation soon gave way to the news that a black geneticist named Rick A. Kittles had launched African Ancestry with his business partner, Gina Paige. Among the earliest direct-to-consumer testing companies in the United States, it was the first niche-marketed to people of African descent. As an ethnographer and historian of African America, with a special interest in science and technology—as befitting a child born from the union of a cryptographer and a mechanical technician—I knew that I had to join Kittles on this journey.

I used what social scientists call "snowball sampling" when conducting my interviews with root-seekers. In other words, I interviewed genealogists about their decision to use genetic ancestry testing and the effects of the results on their lives, and they, in turn, referred me to others. As I would discover, what was snowballing was not only the number of people in my interview network, but the surprising ways the test results were being put to use. That is, I was also being given an unexpected map of how genetic information was being used by individuals, communities, and institutions. Yes, personal and family information was gleaned. But in these conversations there was also growing mention of how broadly genetic ancestry testing was being used as the industry evolved. For over a decade, I've followed Kittles and African Ancestry, and in this time, have come to take a long view of genetic ancestry testing, a perspective that is more mosaic than the predictable, ritualized scenes of revelation and surprise we have become accustomed to witnessing on popular genealogy television shows.

As a wide-eyed girl watching *Roots*, and wondering about mine, I never could have dreamed a future where one day I'd have the surreal experience of having my genealogical results revealed to me before a crowd of African diaspora VIPs and civil rights leaders, and with a prominent actor, Isaiah Washington, as master of ceremonies. Although this experience elicited mixed emotions in me, I can personally attest that new branches on ancestral trees are the undeniable graft of genetic genealogy.

The Social Life of DNA unearths what else we try to accomplish with these tests, including political and legal uses. I've found this might include establishing ties with African ancestral homelands, transforming citizenship, recasting history, or making the case for reparations which, as we know, is an issue that is once again part of our national conversation. I describe these lesser-known but truly momentous uses of genetic ancestry testing as "reconciliation projects," endeavors in which genetic analysis is placed at the center of social unification efforts. These may be legal attempts to financially reconcile formerly opposed parties like the descendants of enslaved persons and the current-day companies that profited from slavery, such as Aetna, JP Morgan Chase, and Wachovia. Reconciliation projects are also efforts to reunite formerly unified parties like blacks in the United States seeking to reconnect with lost kin and community in Africa. They may also be used to reestablish biographical or historical information that has been lost to the march of time or to settle contentious issues. In short, these DNA-based techniques are offering a new tool to examine long-standing issues and these reconciliation projects reveal manifold and potentially transformative possibilities.

The Social Life of DNA tells the compelling, unexpected, and still unfolding story of how genetics came to rest at the center of our collective conversation about the troubled history of race in America. I hope you will join me on this foray into an extraordinary, uncharted arena of twenty-first-century racial politics.

Introduction

The temperature climbed well into the eighties on the day in July 2006 when the cool, damp New England soil was brushed from the top and sides of the coffin of Venture Smith (1729–1805), a formerly enslaved man, under the glaring lights of British Broadcasting Corporation cameras. The BBC was filming a documentary about Smith, titled *A Slave's Story*, in the leafy cemetery adjacent to the First Church of Christ, Congregational, in East Haddam, Connecticut. Joining the film crew at the site were archaeologists, geneticists, anthropologists, and historians. Many of these researchers—whose expertise ranged from African history and colonial America to human genetics and forensic science—were affiliated with two institutions leading the initiative, the University of Connecticut at Storrs, and the Wilberforce Institute for the Study of Slavery and Emancipation in Hull, England. Also present were two senior citizens, Coralynne Henry Jackson and Florence Warmsley, who were among Smith's oldest living descendants. Smith's remains would be examined, with their permission and that of other Smith relatives.[1]

Undisturbed for two centuries, Smith's grave was being excavated because researchers hoped its analysis would illuminate how he lived and died. A team headed by Dr. Nicholas Bellantoni, Connecticut's state archaeologist, was on hand to carry out the excavation. Any genetic samples that could be extracted from the remains would be taken to the laboratory of Warren Perry, a physical anthropologist at Central Connecticut State University, who runs the Archaeology Laboratory for African and African Diaspora Studies. Following this step, a research group led by Linda Strausbaugh, director of the Center for Applied Genetics and Technology at the University of Connecticut, and an expert of "historical genomics," would carry out a genetic analysis of

Smith's mitochondrial DNA (mtDNA), which passes mostly unchanged from mothers to children, for clues into Smith's ancestry.

At a juncture when DNA sleuthing into the past is commonplace and genetic ancestry testing is the stuff of primetime television, this research undertaking seems unexceptional. Yet Smith is a most curious subject for this exercise in excavation and genetic analysis, for the known details of his life rival that of some of our most prominent and well-documented historical figures. Smith recorded a "slave narrative," a first-person account of his life. This narrative is part of a tragic and often heroic genre of American writing that has opened a small window of historical insight onto the experiences of enslaved women and men. Within the genre, Smith's account is distinctive for its fullness. Most uniquely, his narrative goes back in time to his life *before* the Middle Passage, that perilous transatlantic journey that transported Africans into lives in bondage in Europe and the Americas. As historian James Brewer Stewart explains, Smith's narrative "is the only extant account by an African American that links West African memories to a life completed within the United States."[2]

Smith relayed his extraordinary journey from slavery to freedom to Elisha Niles, a local schoolteacher, in 1798, seven years before his death at the age of seventy-six. The account was subsequently published as *A Narrative of the Life and Adventures of Venture, a Native of Africa, but Resident Above Sixty Years in the United States of America, Related by Himself.*[3] In his narrative, Smith goes into fine detail about his life in Africa: he tells us the land of his birth ("Dukandarra, in Guinea"); the name of his father ("Saungm Furro"); his birth name ("Broteer Furro"); and of his experience of capture at the hands of a foreign army led by a man named Baukurre.

Smith recalls that it was Captain Collingwood who helmed the slave ship on which he was transmitted.[4] He shares his story of enslavement from boyhood to adulthood, with his strength and abilities increasing as he grew to be a man more than six feet in stature. Smith also explains how his freedom was secured by his own hand. For, in addition to the duties he carried out as a condition of his enslavement, he did additional backbreaking labor to earn money, and purchased his liberty in 1765. As a free man, he continued to work and save his earnings. Between

1769 and 1775, he paid to liberate his wife, Margaret (Meg), their two sons, Solomon and Cuff, and Hannah, his daughter. Even accounting for the filter of Niles, who transcribed and edited Smith's story utilizing the formulaic conventions of the genre, the former slave left behind a narrative brimming with detail and personal sentiment. According to Stewart, "most historians consider [the narrative to be] a source sufficient in and of itself to explain who Venture Smith was, where he came from, what circumstances he faced, what he believed in, and how he translated his beliefs into designs for living."[5]

Augmenting this rich biographical material, Smith's present-day relatives have extensive knowledge of their ancestry. Smith's descendants have worked with Karl Stofko, the town historian for East Haddam, where Smith was laid to rest, to document ten generations of Venture and Meg's offspring.[6] Additionally, the descendants have a deep and inclusive understanding of family that comprises marriage, partnering, and both formal and informal ("people raised by 'bloodline' descendants") adoption.[7] Smith's descendants have sketched out a tall and broad family tree that begins on the African continent and branches to the present. In fact, the Smith progeny hold the enviable position of having more information about their ancestors than most of us could ever hope to know about our own.

Given that Smith left behind a first-person account of his origins and that we know a great deal about him through historical research and genealogy, what do we make of the scientists, social scientists, and documentary producers who came together in a quest to corroborate his personal narrative with scientific evidence? Why is there a need for scientific support of Smith's life story if historians declare his account a sufficient source in and of itself? Why is DNA analysis deemed to proffer more valuable or reliable information about a man's familial history than his own words?

THE SOCIAL POWER OF DNA

There are two answers to these questions. One concerns the *social power* of DNA. A second, related explanation has to do with the *social life* of DNA. The Smith example illuminates DNA's social power. The quest for genetic evidence suggests a belief that the former slave's true origins are not to be found in his own words or in details provided by his distinctive narrative voice, but lie ultimately in his biological remains. The special status afforded to DNA as the final arbiter of truth of identity is vividly apparent in the language we use to describe it.[8] In a capacity similar to how DNA's bases—adenine, cytosine, guanine, and thymine—combine and recombine in myriad ways to create biological matter, the language of DNA pervades our cultural imagination. Hyperbolic phrases such as "code of codes," "the holy grail," "the blueprint," the human "instruction book," and "the secret of life" suggest a core assumption about the perceived omnipotence of genetics.

In their influential 1994 book, sociologist Dorothy Nelkin and historian M. Susan Lindee discuss these metaphors as part of what they term "the DNA mystique." Drawing on a comprehensive analysis of scholarly and popular evidence, the authors reveal how elements of contemporary culture carry the message that the ultimate source of power and knowledge lies "in our genes." They observe that DNA is seen as a phenomenally malleable symbol, imbued with seemingly magical powers to access the truth.

The gene, they contend, is an icon suffused with "cultural meaning independent of its precise biological properties."[9] While claims that genes tell us who we are or how we will behave are no more true today than they were two decades ago, they are now augmented by the democratization of DNA analysis. In the twenty years since Nelkin and Lindee drew our attention to DNA's authority in the public mind, the gulf between the gene's cultural meaning and its biological properties has narrowed. This is in part the result of direct-to-consumer (DTC) genetic testing kits that now give scientific shape to human yearnings.

The DTC genetic-ancestry-testing industry emerged at the turn of the twenty-first century, growing from a handful of companies to scores of them in the past decade. Reliable figures are hard to come by, but by 2015 the industry was estimated to have served close to two million customers. Half this number came from one company, 23andMe, alone.[10] The explosion of interest in DNA root-seeking in the United States stems from our history as a nation of immigrants and migrants—both voluntary and shackled travelers—arriving from elsewhere and fraying family ties along the way. Tracing one's origins has thus become something of a national pastime.

Genealogy is a pursuit that dates back to biblical times, with the tracing of priestly lines or royal kinship. Beginning in the twentieth century the tracing of ancestry gained widespread appeal in the United States. What sociologist Herbert Gans defined as "symbolic ethnicity"—the ability to hark back to County Cork, Ireland, while jubilating on St. Patrick's Day in Boston, to recall ancestors in Sicily while parading in the Feast of San Gennaro in Manhattan's Little Italy, or to hang an English coat of arms in one's home—became an important component of American identity.[11]

For African Americans, this search is both more elusive and more fraught. A profound loss of social ties was an immediate outcome of the Middle Passage and racial slavery. The ravages of the Civil War left vital records and slave-plantation paperwork degraded or destroyed. Information about black families was also lost intergenerationally—an understandable impulse to forget traumas of the past. But today African Americans use genetic ancestry testing with the hopes of shedding light on precisely the kind of familial and historical information supplied by Venture Smith's slave narrative and family tree.

Thus, the turn to genetic evidence in the Smith case, the endeavor to excavate his remains despite knowing more about him than we do about most Americans of this time period, is revealing. As Smith's descendants have expressed, genetic ancestry testing is about more than the unearthing of facts. They hold, and we hold also, more intangible aspirations for DNA.

Since 1997, the community of East Haddam has celebrated Venture Smith Day, an occasion that typically includes scholars like the historians John Wood Sweet, an expert on the life of the former slave; Robert Hall, who both studies Smith and has performed as him in period dress; and Chandler Saint, president of the Beecher House Center for the Study of Equal Rights; and local politicians such as Connecticut state representative Melissa Ziobron. This day also marks the annual Smith family reunion, when descendants flock to Connecticut to lay a wreath on their patriarch's gravesite, take a group photograph, and commune with one another.

When I encountered Florence Warmsley, an eighth-generation descendant, at the tenth annual Venture Smith Day in September 2006, she told me she hoped that the genetic testing of her forefather would help to "educate the American public." Her elder cousin Coralynne Henry Jackson, also an eighth-generation descendant, felt similarly about the educational importance of DNA analysis of Smith's remains. In *A Slave's Story*—which premiered in the spring of 2007 on the bicentennial anniversary of the abolition of the British slave trade—Jackson declares her hope that the results of the investigation into Smith's remains would "help schoolchildren . . . [and] let us know where in Africa he came from."[12] To be sure, Jackson and Warmsley want to uncover details about their ancestral past. Yet notably, both also express a desire for the education of *others* (schoolchildren, the American public, and so forth). They hope that this excavation work, in unearthing scientific data that might teach us something more about Smith's personal history, will also yield information about the history of slavery and racism in the United States.

As *A Slave's Story* closes, the camera fixes on Warmsley, who poignantly proclaims her hope that the revelation of her ancestor's genetic ancestry will "bring healing."[13] The desire expressed by this distant descendant of an enslaved man and woman gets to the very heart of what some African Americans hope can be accomplished with genetic genealogy testing: racial reconciliation.

Genetic analysis is indeed increasingly being used as a catalyst for reconciliation—to restore lineages, families, and knowledge of the past and to make political claims in the present. We now turn regularly to genetic testing to unravel mysteries and resolve questions. As I describe in this book, the social power of DNA is being similarly leveraged to raise awareness of blacks' past experiences and, in so doing, contribute to today's racial politics, which are too often marked by historical amnesia.

I came to this conclusion reservedly. I believe that the ultimate account of Venture Smith's origins is not to be found in his genes, the most essentialist and socially anemic conception of human identity. I am wary when genetic inference is given trump power over historical archives and autobiography. Moreover, the excavation of Smith's grave for DNA analysis is more telling of how our ideas about race are unfolding in the genomics era than it is about Smith's authentic self. At the same time, the Smith case is invaluable in showing that part of the appeal of genetic ancestry testing lies in providing a lexicon with which to continue to speak about the unfinished business of slavery and its lasting shadows: racial discrimination and economic inequality. DNA-based techniques allow us to try—or try again—to contemplate, respond to, and resolve enduring social wounds. Today, genetic science "can play a role in many different stories," as Nelkin and Lindee put it, as it moves beyond the hospital, the clinical laboratory, and the courtroom to the work of reconciliation, swinging like a pendulum between our hopes and fears.[14]

THE SOCIAL LIFE OF DNA

The summer of 2010 marked the tenth anniversary of the decoding of the "first draft" of the human genome. This scientific milestone was met with cautious appraisal from usually enthusiastic quarters. Leading science reporter Nicholas Wade bemoaned in the *New York Times* that "after 10 years of effort" the therapeutic promise of genomics "remains largely elusive." J. Craig Venter, the maverick geneticist who was a major force behind the Human Genome Project, declared more emphatically in *Der Spiegel* that "we have learned nothing from the genome." These sober assessments were surprising given how profoundly our social world has been changed by genetic science and its applications in the last decade.[15]

While concrete health benefits stemming from the Human Genome Project may indeed be "elusive," its broader impacts are clear. Genetic analysis has become widespread. There is no question that genetics research is prevalent in biomedicine, even if its ability to predict or remedy ills remains to be fully demonstrated. In criminal justice settings, DNA is becoming ubiquitous and is double-edged, playing a role in both conviction and exoneration. And commercially available genetic tests that claim to specify genealogy, ancestral affiliation, or racial and ethnic identity are among the most conspicuous signposts of our genome age. In these different institutional settings, we have zealously (and often uncritically) seized upon DNA as a master key that unlocks many secrets.

DNA is the ultimate big data. Genetic data is multivocal and contains information that can be used in varied facets of society regardless of its source or its original intent of use. And genes are omnibus; they confer many types of information simultaneously. DNA analysis therefore moves across and between the expected medical, forensic, and genealogical domains, and also beyond them into a wider set of arenas, with expanded purpose. Diverse ends and aspirations are now sought with and through the use of genetics. This diffusion comprises the social life of DNA.

This social-life approach to genetics follows the methodology of the anthropologist Arjun Appadurai, who, in *The Social Life of Things*, argues that to understand what objects mean and why they are important we must trace their circulation in society ("things-in-motion"), illuminating "their human and social context" and revealing "the human transactions and calculations that enliven things."[16] By similarly tracking DNA analysis, we gain insight into where and why genetics is called upon to answer fundamental questions about human existence, often through extensions of its popular genealogical uses. Genetics is today engaged in practices of identity formation, in philanthropy and socioeconomic development projects, as corroborating evidence in civil litigation and historical debates, and elsewhere. Thus, although the therapeutic utility of the genome may be arguable, the social life of DNA is unmistakable: the double helix now lies at the center of some of the most significant issues of our time.[17]

RECONCILIATION AS SOCIAL PRACTICE

Reconciliation projects employing genetics are illustrative of the social utility of DNA.[18] In these endeavors, genetic analysis is used to contribute to community cohesion, collective memory, or social transformation. This is the very same "healing" that the Smith descendants sought for themselves and for American society. With reconciliation projects, DNA analysis is incorporated into attempts to reunite formerly opposed parties or formerly unified ones (rejoining broken ties within a family, a community, a nation-state, or a diaspora); to uncover biographical or historical information that has been lost to the march of time; or to adjudicate contentious issues. Reconciliation projects are forms of social practice that happen across the globe. In post-conflict Argentina, in the well-known case of the Asociación Civil Abuelas de Plaza de Mayo (Grandmothers of the Plaza de Mayo), for example, genetic analysis has been used for more than two decades to help reunite children—who lost their parents to political violence and were unlawfully placed with adoptive families—with their biological grandparents. In post-apartheid South Africa, DNA analysis has helped to identify the bodies of former members of the African National Congress who were "disappeared" in the fight against state-sanctioned racial discrimination. What these efforts share is the use of forensic evidence to elicit information about the past, often after conflict or trauma.

This book considers one such project in depth: efforts aimed at repairing the social ruptures produced by transatlantic slavery. I examine a constellation of activities initiated by black Americans and other persons of African descent, who embark upon genetic journeys of discovery about their ancestral origins and who, in turn, propel this information into a variety of sociopolitical purposes. With reconciliation projects, DNA analysis has been annexed onto unresolved and, therefore, persistent debates about national belonging. Beginning with the emergence of "humanitarian genetics" in Latin America and moving forward to the "founding family" controversies in the United States, developments in the 1990s combined to make possible a new course for molecular biology—one set toward reconciliation.

I began my research for this book in 2003, when genetic ancestry testing was in its infancy. Attending genealogical conferences and events throughout the United States, I encountered the early adopters of these techniques, amateur researchers who had been targeted by purveyors of genetic-testing services because of their demonstrated interest in reconstructing family history. These gatherings attracted a rich community of African American genealogists.

The Social Life of DNA draws upon my encounters with persons of African descent who have used genetic genealogy testing. For more than a decade, I have observed and participated in occasions during which genetic genealogy testing was discussed or offered, including meetings at churches, community centers, libraries, museums, and universities in both the United States and the United Kingdom. I conducted fieldwork and interviews in a wide variety of settings including at local and national gatherings of genealogists, at African American heritage tourism sites, and in living rooms and at kitchen tables.

The practice of genealogy now involves a wide array of technical tools alongside traditional archival research: Ancestry.com's Family Tree Maker and other computer software programs assist genealogists in constructing pedigree charts; there are e-mail listservs dedicated to the technical aspects of genetic genealogy testing; and there are websites at which test-takers can compare DNA results in order to establish degrees of relation. Because of the technological innovations in genealogical practice, my research necessarily involved "virtual" ethnography via the Internet.[19] Specifically, I observed and participated in virtual communities of people who shared an interest in tracing their African ancestry.

Online community discussions encompass varied topics related to the practice of genealogy. For instance, one forum is dedicated to discussions of DNA testing. In these settings, members dialogue about the science behind genetic ancestry tracing; display expertise through their command of jargon and recent genetics research, or developed through prior experience with one or more testing companies; circulate topical scientific papers and newspaper articles; and share genetic genealogy test results and their feelings about them. In more recent years, I followed the emergence of videos by genetic genealogists that broadcast their desire for roots and their genetic ancestry test results to a social media audience. My involvement in these online communities consisted of discussions with genealogists, both on the public listserv and "off channel"—that is, in private online conversations. In most cases, there were points of overlap and continuity between the online and offline communities I inhabited and, as I describe, I came to know members of this virtual community personally through interviews and at genealogical gatherings.

A preponderance of the black root-seekers I encountered early in my exploration of genetic genealogy used the services of the pioneering African Ancestry company (either exclusively or in combination with other companies' tests). So, in addition to gleaning insights from genealogists, I began to track the evolution of this particular DTC genetic-testing enterprise. Founded in 2003 by Rick A. Kittles, an African American geneticist now based at the University of Arizona, and Gina M. Paige, a black businesswoman who holds a bachelor's degree in economics from Stanford University and completed an MBA at the University of Michigan, African Ancestry traces maternal and paternal lineages, determining an individual's affiliation with nation-states and ethnic groups on the African continent.[20] (Although, by necessity, the company also provides genetic-ancestry-testing matches to sites outside the continent of Africa.)

In the course of completing my research, I became a member of the Afro-American Historical and Genealogical Society (AAHGS), frequenting a local AAHGS chapter in the Harlem neighborhood of New York City. I also interacted with genetic genealogists and eventually became one, purchasing a genetic genealogy test from African Ancestry and participating in a live "reveal" of my test results.

The transformative role that Kittles and Paige's genetic genealogy testing can play has been powerfully depicted in Henry Louis Gates Jr.'s popular television series *Finding Your Roots*, several segments of which relied upon African Ancestry's products and dramatic on-camera "reveals." In this book, I uncover some of the even larger-scale transformations that are taking place as genetic analysis moves beyond the bounds of clinical research laboratories and criminal justice courtrooms and genetic genealogy takes on more than solely personal significance.

Back in 2000, rumors that DNA testing might soon be available to aid black root-seekers were confirmed in the *Los Angeles Times*. Kittles, then a codirector of the molecular genetics unit of the National Human Genome Center at historically black Howard University, was quoted as saying that "hundreds of African Americans" had sought him out based on the mere suggestion that he was months away from releasing a genetic test that "could link them to their long-lost lineage."[21] These were heady, millennial days, and as a sociologist of science and of race, I was brimming with curiosity about what brave new world might be ushered in with technologies that the *Times*' reporter said "sound[ed] like something out of an episode of 'Star Trek.'"[22]

At first, I wondered if these innovations, such as the genetic genealogy test then in development by Kittles, would spur changes in our conceptions of racial identity. How might the notions of race and ethnicity be transformed after the decoding of the genome? Did these tests spell the end of entrenched racial ideologies of difference and the beginning of new possibilities for human identity?

"OUR COMMON HUMANITY"?

At the time I began my research, the literature on genetics and society was small, but there was concurrence on this point: developments in contemporary DNA analysis, especially genetic ancestry testing, threatened to retread the tragic path of scientific racism. By the late 1990s, scholars, including the previously mentioned Nelkin and Lindee, Barbara Katz Rothman, and Troy Duster, warned that without a course correction, we were in danger of becoming a society ruled by genetic determinism—a dystopian future world in which one's biological inheritance was believed to indelibly shape one's health, behavior, and other attributes.[23]

Several explanations have been offered for this prediction. Rothman lamented the creation of a "gene for everything" society in which DNA was understood to be the alpha and omega of human life. She also pointed to the eugenics of the twentieth century as a cautionary tale. Duster contended that eugenics emerged from, bolstered, and justified racist ideologies that had never really gone away and were reentering American society through the "back door," disguised as concern for greater knowledge or better health for all. As we crossed the threshold into the genome age, distant and more recent histories of scientific racism elicited words of caution from social analysts.

In the media and throughout the public sphere, although some reservations were aired, the "new genetics" was mostly framed as a silver bullet. This framing was at its zenith during the announcement of the decoding of the preliminary draft of the human genome at the White House in June 2000. A central conceit of the Human Genome Project was that we could derive a great deal of information—indeed, life's ultimate data—by deciphering the genetic signatures of a select, unidentified and multicultural group of five persons, three women and two men. (This group included Celera Genomics founder Venter, who spearheaded the private arm of the ambitious research endeavor.)[24] Because, with "the goal of achieving diversity," the human genome is a composite of different individuals, the logic went, it promised an end to the ideologies and hierarchies of human difference that had once justified bondage and continue to stoke interracial conflict and inequality.[25]

Under the flashing glare of the international press corps' cameras, President Bill Clinton—flanked by Venter and National Human Genome Research Institute director Francis Collins, and with Prime Minister Tony Blair of Britain joining in via remote video—breathlessly declared that "one of the great truths to emerge from this triumphant expedition inside the human genome is that in genetic terms, all human beings, regardless of race, are more than 99.9 percent the same . . . modern science has confirmed what we first learned from ancient fates. The most important fact of life on this Earth is our common humanity."[26] In effect, several of the world's leading political and scientific figures bore witness to our fundamental human sameness. The message: composed of an amalgam of individuals' genes, the human genome represents us all. The bilateral press conference featuring heads of state Clinton and Blair revealed moreover that this scientific project was also a political one: from *e pluribus unum* to *e unum pluribus.*

But the powerful supercomputing technologies that enabled the unraveling of the genome were swiftly put to the task of parsing that sliver of difference—one-tenth of a percent or less—between humans. And the difference that was deemed to matter was race. As sociologist Dorothy Roberts put it in a recent Human Genome Project postmortem, "reports of the demise of race as a biological category were premature."[27] Moreover, she harked back to the warnings of social scientists like Rothman and Duster, writing that "biological theories of race" are being resuscitated through the use of "cutting-edge genomic research" that "modernize[s] old racial typologies."[28] Indeed, soon after the genome was mapped, both science and industry quickly turned their attention to the mining of that thin vein of genetic variation between individuals and groups, in the process presenting the possibility of resurrecting dangerous racial hierarchies, and not through the back door, as Duster augured, but through a sideways proxy discourse of ancestry, genealogy, geography, and population.

Within months of the completion of the draft of the human genome, leading genetic scientists published controversial papers concluding that humans can be classified into groupings that confirm the biological reality of race. These claims in turn opened the door to the reanimation of discredited typologies that have existed since the time of Carl Linnaeus. Linnaeus, the eighteenth-century Swedish taxonomist, extended his classification of plants and animals to human beings. He contended that our human community, *Homo sapiens*, is composed of four racial subcategories, or taxa, in this order: Europeanus, Americanus, Asiaticus, and Africanus.[29] But unlike in his cataloging of flora and fauna, with humans Linnaeus linked appearance and social behavior, noting, for instance, that those in the category Europeanus are "white . . . very sharp, inventive," while those under the banner Americanus were characterized as "red, ill tempered, subjugated." DNA-age racial claims rely on techniques that allow researchers to locate genetic variants shared across human groups but differently distributed among them. Extending from a Linnaean taxonomy that linked skin color and behavior and arranged these hierarchically, some then attribute racial and ethnic distinctions to this statistical spectrum. Some of these researchers argue that it is the nation's pressing healthcare needs that oblige them to use racial and ethnic categories as "starting points" for their research.[30] Invoking scientific objectivity, others deny any responsibility for how their research findings might be used to bolster racist claims.

For his part, the former *New York Times* science reporter Nicholas Wade did not hew to scientific objectivity when he published an incendiary book—*A Troublesome Inheritance*—that threw into question any uncritical celebration of the socially transformative possibilities of DNA analysis.[31] Wade misread and then misused genetics research—including the research mentioned above—to argue that there are three biologically distinct human races (African, East Asian, and Caucasian) that in turn have similarly distinct, inviolable social behaviors. According to Wade, these alleged differences account for disparities in human societies. Wade attributes the "rise of the West" to Caucasians' superior genes, which yielded trustworthy, entrepreneurial societies, while African genes bred cultures of violence and distrust, and East Asian genes yielded practices of rigid discipline and stratification. The genetic inheritance of these latter groups, he argues, left them incapable of the social cohesion and advances of "Western civilization." Wade's book was roundly discredited as obtuse on the science and dead-wrong with respect to its implications in a letter to the *New York Times* signed by over a hundred scholars with expertise in genetics, human biology, biological anthropology, and evolution. But its welcome reception in other, less-expert quarters showed how quickly new life could be breathed into long-disproven racial stereotypes.

GENETIC POLITICS

However, the obvious risks posed by willfully speculative and ideological work such as this should not hinder necessary conversations about the politics of race after the genome. Discussions of race and ethnicity in the genomics era need to be pursued from many perspectives, and these inquiries must take into account the racial dynamics surrounding the human genome. The genome emerged at a specific political moment, at a time when a form of racial ideology that sociologist Eduardo Bonilla-Silva has termed "color-blind racism" was coming into strength.[32] Thus an odd paradox seemed to be at hand: race was "becoming more significant at the molecular level" at the precise time it was being declared "less significant in society."[33]

Genetic politics is not a politics like any other. There are properties unique to DNA that deem it suitable for making political claims. DNA can be used to embody the past, and because DNA is shared it can represent both individuals and groups. DNA can be used to highlight a history of oppression that has been rendered invisible. Black Americans are using DNA analysis to advance issues important to them in a political climate that is increasingly indifferent to demands for social and racial justice. As sociologist Stephanie Greenlea contends, today's racial justice movements "must now attend to these tasks in a context where erasures and silences on racism threaten to render the very basis of complaint invisible."[34] In short, combating color-blind racism requires the restoration of color-*vision*—that is, the return to visibility of historic and continued racial inequalities. Genetic ancestry testing is being used to make this case.

In this "post-racial," post-genomic moment, therefore, DNA further offers the unique and somewhat paradoxical possibility of magnifying issues of inequality in order to bring them into view, both literally and figuratively. Social inequities may then be challenged using other strategies such as the courts and social movements.[35]

Greenlea demonstrates this phenomenon by analyzing the experience of the Jena Six—six African American youth in Jena, Louisiana, who, in 2006, were excessively charged, tried as adults, and convicted for assaulting a white classmate. Initially, the public framing of the case had the six teenagers attacking their peer without cause. But activist efforts highlighted the racist harassment the boys had been subjected to: they were responding to the hanging of a noose from a tree on campus. The incident was transformed into an activist cause célèbre when the history of this symbol was brought into view by Jena Six supporters in mainstream and social media (not unlike the debates over the meaning of the Confederate flag after the tragic June 2015 murders of nine black church members by a white supremacist in Charleston, South Carolina). Genetic ancestry testing can similarly effect a reckoning with history. Like the rope with a slipknot, genes do not have inherent meaning outside a social and political context. But both signify so much.

At the same time, owing to the transitive property of DNA, political claims made via genetics are always simultaneously about the individual and a collective. An individual's DNA contains genetic information about their biological parents and extended biological family. Native American communities are keenly aware that one person's DNA overlaps with that of many others, and also that their interests are not always served by science. In an episode of Gates's PBS series *Faces of America* airing in the spring of 2010, indigenous author Louise Erdrich consults her tribal elders about her invitation to undergo genetic ancestry testing. Erdrich ultimately refuses to participate in the testing because she and her community understand her DNA to be communal property. As bioethicist Dena S. Davis writes, DNA "extend[s] beyond individual subjects. Genetic research can also affect the communities to which the subjects belong, by rewriting the narratives and reconfiguring the identities that members of the community share and live by."[36] Accordingly, reconciliation projects that appear to deal with one individual's family history practically and symbolically always concern a larger body politic.

Many are attuned to the historical danger of genetics and to the scientific flattening of human society such as the issues that Erdrich and her community weighed. So why turn to DNA for leverage in racial politics, given the fraught history of scientific racism from Linnaeus to the present? Political actors engaging genetics in their political claims-making are not necessarily guilty of self-delusion. Rather, it is precisely because they have a deep appreciation for the complexity of race as a political category that they are mobilizing it as such.

Although I initially set out to understand the extent to which race was being reconfigured from a social and political category to a biological one, almost immediately my research turned to root-seekers and their use of genetics in practice. This focus established that concerns about racial reification and genetic determinism, while well warranted, did not begin to address the myriad purposes to which genetic ancestry is being put. The call for an exercise of caution with respect to political claims based on DNA does not translate into opposition to the use of genetics for the purposes of reconciliation. Indeed, I believe we must remain open to the techniques of the digital and genomic age as they are being applied to political and social efforts.

FOUNDING FAMILIES

By the early 1990s, reconciliation projects utilizing DNA were becoming prominent. The effort in Argentina to reunite families in the wake of political violence was internationally known.[37] In eastern Europe, an analysis that established the identity of several members of the Romanov family of czarist Russia gave a boost to the use of DNA in historical inquiry.[38] In the United States, some efforts ignited controversy stemming from the country's history of imperialism, slavery, and dispossession. DNA analysis was initiated on the nine-thousand-year-old remains of a Native American man (the "Ancient One") uncovered near Kennewick, Washington, in 1996, for example, despite vehement resistance from indigenous groups in the Pacific Northwest who demanded that the remains be returned and reburied under the Native American Graves

Protection and Repatriation Act (NAGPRA).[39] The benefit of the research was hotly debated, raising the question of whether it would serve only to compound historical exploitation. Many Native Americans, who held no uncertainty about the Ancient One's origins, surely believed so. (In 2015, a study of the Ancient One's genome undertaken with the permission and cooperation of the Confederated Tribes of the Colville Reservation corroborated the indigenous community's claims.)[40]

Around the same time, genetic answers were beginning to be pursued to questions surrounding the American founding founders. Was Thomas Jefferson father to the children of Sally Hemings, the enslaved woman owned by him? Widely acclaimed research by the legal historian Annette Gordon-Reed contends that both the norms of plantation society and a preponderance of archival evidence—the scholarly gold standard—substantiate Jefferson's paternity of at least one of the Hemings children.[41] Nonetheless, scholars furiously debated this question and resistance to Gordon-Reed's findings remains stubborn in some camps. In the late 1990s, researchers embarked upon a genetic investigation to settle the matter. It is important to note that DNA analysis in the Hemings-Jefferson case was one of the earliest instances of these techniques being used in the United States. Not only was the effort controversial and technically pathbreaking, it opened the door to a new strategy for trying to come to terms with the history of slavery and its contemporary effects, in the face of resistance to a full airing of America's racial history.

In the investigation, genes of Jefferson's avowed descendants were compared to those of his purported ones. This analysis drew on the distinctive features of *Y-chromosome DNA* (or Y-DNA) to determine whether there were familial links. Y-DNA is passed mostly unchanged from fathers to sons—women do not have a Y chromosome—and can be used to trace a direct line of male ancestors, including uncles and nephews. In this case, the Y-DNA of male descendants of Jefferson's paternal uncle was compared to that of two of Hemings's sons, her first and her last. The resulting paper, published in 1998 in the leading science journal *Nature*, declared in its very title that "Jefferson fathered [the] slave's last child."[42] This finding was based on comparison of male-line genetic markers. Male descendants of Hemings's youngest son, Euston, were revealed to share a Y-chromosomal *haplotype* with the male relatives of Jefferson. (Haplotypes are sets of DNA variations that are typically inherited together.) The report noted that this genetic marker was so rare that "it has never been observed outside of the Jefferson family."[43] Despite geneticists' claim—and historians' concurrence—that Jefferson's paternity of his slave's youngest son is the "simplest and most probable explanation," some of the late president's living descendants, who argue that it was Jefferson's brother who fathered the Hemings child, continue to vehemently refute this claim.[44]

The Jefferson-Hemings case was the first time that DNA evidence would be called upon to confirm a slave descendant's shared ancestry with a founding father. But it would not be the last. A few years later, alleged descendants of James Madison, the "Father of the Constitution," who followed Jefferson into the US presidency, endeavored to use a similar strategy to prove their place in the Madison clan and in the nation's history. In 2004, Bettye Kearse, an African American physician, joined forces with geneticists at the University of Massachusetts, a team led by black geneticist Bruce A. Jackson, in the hopes of establishing her family's place in the Madison lineage.[45]

Madison and his wife, Dolley, had no children together. But generations of oral history in Kearse's family held that Madison did have a son. As an article in the *Washington Post* elaborated, this history "begins with a kidnapped African slave, Mandy, who Kearse says was impregnated . . . by Madison's father. The child, Coreen, later gave birth to Madison's child." Madison's purported child was Kearse's great-great-great-grandfather.

As with the Jefferson case, Y-chromosome analysis was to be used to compare the DNA of one of Kearse's male relatives to that of Madison's descendants. But because this known male progeny of Madison denied her request for a DNA sample, Kearse's kinship claim remains unconfirmed by science, even if it has been long cemented in her family's oral history.[46]

The chosen method of resolution—DNA analysis—places these examples firmly within that category of social initiatives termed reconciliation projects. Information gleaned from DNA is used to establish social inclusion or exclusion, mediate social justice claims, or resolve sociohistorical and political controversies. At stake as much as the genetic truth of the matter is what such DNA histories would reveal about the hypocrisy in play at the time of the nation's creation. Founding fathers who wrote passionately about freedom both held slaves and likely bore children with women who were deemed biologically inferior (Jefferson) and not worthy of personal liberty (Madison). A fuller appreciation for and verification of this history could reshape the foundational mythologies of American society and reshuffle our collective memory of our first families. These efforts to right the nation's founding narrative heralded the incorporation of genetic identification techniques into American racial politics and, more specifically, into considerations of the history of slavery.

Today genetic analysis is being used as well in an array of less widely publicized efforts that seek to document the transatlantic slave trade and address its devastating social consequences. More specifically, the tests proffered by the African Ancestry company, MatriClan and PatriClan, are promoted as a means to uncover details about the experiences of enslaved blacks; to redeem and restore today's stigmatized African American family by reuniting it with an ancient, idyllic "African" one; to cement affiliations among members of a transnational network of blacks; and to resolve the social injury inflicted by the "peculiar institution." In all of these enterprises, cofounder Rick Kittles's role as a scientific entrepreneur cum cultural advocate is as crucial as the genetic tests he developed. *The Social Life of DNA* follows Kittles and his company's DTC genetic tests as they traverse the landscape of our so-called "post-racial" society.

THE CHAPTERS AHEAD

The next chapter of this book discusses the global emergence of reconciliation projects incorporating DNA analysis. A broad set of practices, reconciliation projects involve the use of a forensic or truth-telling mechanism—be it human testimony or scientific data—to probe the past. I introduce two milestone events that prefigured this possibility. First is the landmark use of "genetic technologies as tools for human rights" by Las Abuelas de Plaza de Mayo (the Grandmothers of the Plaza de Mayo) in order to locate the grandchildren of their sons and daughters who were "disappeared" by the junta state during the Argentinean "Dirty War" that spanned from the late 1970s to the early 1980s. Second, I describe a more topically related development, the United States' first-ever truth and reconciliation process, which was inaugurated in South Carolina in 1999, on the twentieth anniversary of brutal, race-based murders in the city of Greensboro. Here the testimony of both victims and perpetrators facilitated the collective airing of festering tensions. The ways that African Ancestry's genetic genealogy tests are taken up borrow elements from both of these earlier efforts at social healing: innovative uses of DNA analysis on the one hand, and confronting and drawing lessons from past transgressions, on the other. The confluence of enduring social harms and new technologies brought us to a historical moment at which genetic ancestry testing would take on a broader than intended purpose.

We look at a research study initiated around the African Burial Ground in New York City in the early 1990s in chapter 2. Here important technical cornerstones were laid for future genealogical pursuits via genetics. The rediscovery of this colonial-era burial ground, with its promise of rare insight into the life and death of bondspersons in New York, was an occurrence of great historical import. It would also become a signal event in the evolution of direct-to-consumer DNA testing, as discussed in chapter 3. The political and scientific debates that arose around the issue of the disposition of this landmark cemetery—now a property of the National Park Service and renamed the African Burial Ground National Monument—was foundational to the proliferation of genetic analysis and its uses in black cultural politics.

Rick Kittles began his career as a junior scientist working on the study of this historic site, and the idea for African Ancestry—the company he would cofound several years later—was conceived as a result of this work. Equally important, black community activists laid a symbolic foundation for the reconciliation projects described herein. An activist group that called itself Descendants of the African Burial Ground steered the course of the genetic research at the cemetery in such a way as to restore African "ethnicity" to the unknown blacks buried in it. In so doing, they also anticipated how DNA analysis would be marshaled for the collective betterment of persons of African descent, above and beyond the intimate, personal quests that we have now come to most closely associate with genetic ancestry testing.

How individual genetic data is transformed into collective cultural politics is the centerpiece of this book. Exploring how genetic genealogists of African descent put their test results into action, transforming their individual roots pursuits into reconciliation projects, is the subject of chapter 4. I describe ethnographic encounters with Kittles, his business partner Paige, and their customers. For Kittles, African Ancestry is not simply a business enterprise. Following him across the United States, we observe how he links his work to a broader racial justice vision that imagines the liberation of black communities through ancestral knowledge. While many of Kittles's customers come to African Ancestry seeking basic genealogical information, in my conversations with them, they spoke of the desire to feel complete, of craving a stronger sense of belonging in the United States and on the continent of Africa, and of wanting in their own way to reckon with the history of slavery—all of this in addition to the inferences they receive about the ethnic groups in Africa to which they have been linked.

Inspired by Alex Haley's roots journey of decades ago and by more recent television depictions of root-seeking, black genealogists seek out genetic testing to cultivate their family trees. Unlike the African Burial Ground study, however, this genetic analysis is carried out on the DNA samples of living persons. These genealogists can therefore actively fashion new identities from the genetic cloth supplied by African Ancestry. They then strategically shape their genetic results into "usable" pasts that open new avenues of social interaction, including travel and civic engagement. Although this root-seeking is often animated by blacks' yearnings for pre-slavery identity, given the always communal nature of genetic information, these results may also be scaled up and used to make claims to and for DNA kin. The myriad ways that genetic ancestry tests travel is the subject of chapters 4 and 5.

In chapter 5, we explore the role of the genetic genealogy "reveal." For African Ancestry, the public reveals of black celebrities' test results began in 2003 as a marketing strategy. Within a few years, the practice became a narrative element in genetic genealogy reality television. The reveal is now an accepted stage in the genetic ancestry-testing journey, so much so that root-seekers today use social media to reproduce and perform this key moment in the arc of their experience. The reveal also reminds us that the work of reconciliation for which genetic genealogy may be used includes a larger audience to bear witness to it. As such, the reveal can be a political occasion that asks viewers to take note of the historical circumstances that make genetic ancestry testing important to the root-seeker.

We meet a multifaceted group of Sierra Leoneans that includes the actor Isaiah Washington. He learned of his genetic association with Sierra Leone during one of African Ancestry's early reveal ceremonies. The test results inspired his involvement with a series of activities that drew him closer to the country and its diaspora. Washington was on hand on a winter morning in 2009 when I arrived in Charleston, South Carolina, to attend "a ceremony of remembrance" of ancestors dispersed or lost by the slave trade. The majority of the people gathered that day laid claim to Sierra Leone in some manner, as did Washington, a self-described "DNA Sierra Leonean." The ceremony was organized by participants of "homecoming" trips from the United States to Sierra Leone that had taken place beginning in 1989. During these three pilgrimages, Sierra Leonean elders performed ceremonial rites in which they summoned the "common ancestors" of the Americans and the Africans to "bless their homecomings and bring their broken family back together." Genetic identity was a vehicle to a wider territory—the terrain of reconciliation.

Chapters 6 and 7 consider the role of Kittles's genetic ancestry analysis in a historic class-action suit for financial restitution for unpaid slave labor, a case that originated in a Brooklyn federal court in 2002. Led by activist-attorney Deadria Farmer-Paellmann—a founder and executive director of the New York–based reparations rights organizations the Restitution Study Group and the Organization of Tribal Unity—the plaintiffs availed themselves of genetic genealogy testing carried out by African Ancestry to advance their case. In this effort, DNA test results were submitted as evidence of their forbearers' enslavement and subsequent loss of both ancestral identity and wealth. Here reconciliation took the form of restoring inheritances of community and capital.

The formation of "DNA diasporas" or "DNA networks" is elaborated in chapter 8. Spearheaded by organizations such as the Leon H. Sullivan Foundation, an organization with deep roots in the civil rights movement, as well as by entrepreneurial individuals such as Isaiah Washington, DNA diaspora networks are efforts to foster "reunion" and collaboration among a transnational group of blacks through philanthropy, economic development ventures, dual-citizenship schemes, and heritage tourism. African Ancestry's DNA testing service is being employed as a basis of affiliation for networks forged between continental Africans and blacks abroad. These initiatives resemble the historical "back to Africa" and Pan-Africanist enterprises of Marcus Garvey and others but, importantly, they point as well in a new direction, toward a possible future of genetically derived global racial politics.

However, there are also complications that accompany this dream of transnational reconciliation, as a controversy involving Washington made clear during the 2014 Ebola outbreak. Washington has dual citizenship with Sierra Leone, one of the countries worst hit by the epidemic. *Sierra Express Media*, a local newspaper, criticized Washington for carrying a passport for the country without sharing its burdens during this health emergency. He responded "as a citizen of Sierra Leone," establishing a fundraising effort to support the eradication of the disease through his philanthropy, the Gondobay Manga Foundation, and by lobbying at the United Nations in 2014 for intergovernmental support to help combat the epidemic. As this example suggests, DNA diasporas create networks of privilege *and* obligation.[47] The Sullivan Foundation had a "grand vision" for bridge building between the African continent and its diaspora. Yet this proposal also raised the issue of what kinds of relationships, including kinship arrangements and citizenship, were warranted based on genetic ancestry testing results.

This book follows a collection of interrelated reconciliation projects from their beginnings in the United States in the 1990s to their transnational expansion today as they have evolved from individual pursuits to a wide array of sociopolitical undertakings. Tracking these reconciliation projects allows us to perceive how desires for reunion and recompense—alongside the expansion of direct-to-consumer DNA testing—have pushed genetics beyond the small social world of the biomedical laboratory into larger society.

In light of African Ancestry's burgeoning business, what does the use of genetics in the context of long-standing cultural aspirations and political struggles suggest about the prospects for racial reconciliation in the United States? Activist groups and NGOs—including some, like the Leon Sullivan Foundation, with direct ties to the late-twentieth-century civil rights movement, and others, like the Descendants of the African Burial Ground, that carry the march toward racial justice forward into the twenty-first century—are turning to genetics to accomplish goals formerly pursued through grassroots organizing, electoral politics, and moral suasion. Laments about the decline of the civil rights activist tradition may thus be misplaced. Activism is simply taking new forms, in line with the scientific and technological innovations of the last decade, as the social media campaigns that sparked national protests against police brutality in Jena, Louisiana; Ferguson, Missouri; Baltimore, Maryland; and at other sites in recent years make clear. And given cautionary observations about the lack of participation of minorities in science and technology and these same communities' traditional (and well-founded) mistrust of these and related fields, the embrace of genetics for liberatory ends is particularly striking.

However, efforts to reclaim original identity through genetic technologies, while psychically beneficial, fail to materially address persistent structural inequality. In the same way that the scientific and therapeutic applications of DNA research may not yet have fully lived up to expectations, the application of genetic technologies to reconciliation initiatives brings their technical, institutional, and political limits into stark relief. The trajectory of the reconciliation projects discussed in this book suggests that ultimately, equality, justice, and ethics are not easily tethered to or readily settled with DNA evidence.

We should worry that, with their reliance on commercial products, well-intentioned, innovative uses of genetic genealogy might contribute to a world in which claims for citizenship are tied to practices of consumption. And we should worry mightily about the transposition of the principles of justice into scientific techniques. But we should also appreciate that these endeavors are an innovative strategy on the part of some who find other avenues to historical awareness and social justice blocked, and who pursue the road to racial reconciliation nevertheless.

ONE

Reconciliation Projects

> There comes a time in the life of every community when it must look humbly and seriously into its past in order to provide the best possible foundation for moving into a future based on healing and hope.

Charting the social life of DNA, we discover that forms of genetic testing considered "recreational"—such as those used to infer an individual's ancestral origins—can have as great an impact on society as testing in seemingly more weighty sectors like medicine. In tracking a social path for genetics, our attention shifts from a limited focus on the individual to the broad spectrum of ambitions surrounding these tests. Among these ambitions is the effort to make "whole what has been smashed."[1] Put another way, DNA has become an agent in the politics of repair and reconciliation; it is sought after as communal balm and social glue, as a burden of proof and a bridge across time. Though even a molecule as elegantly complex as DNA cannot possibly fulfill all these expectations, its recruitment into unanticipated social and political uses is both fascinating and telling.

Reconciliation projects epitomize the expansive sociopolitical use of genetics that is the focus of this book. These activities include reparations activism and the quest for acknowledgment of a social injury. They can be campaigns for state apologies following internecine conflict or the uncovering of political atrocities. Public deliberations that have as a goal a new collaborative understanding of the past also fall into this category. So too does the work of bringing together estranged communities.

What these diverse missions share is a forensic mechanism for getting at "the truth." "Forensic" here relates to both legal discussion or public debate and scientific evidence used to support a claim. Whether witness is borne through testimony and narrated memory as in formal truth and reconciliation processes or acquired through genetic data, forensic information is used in attempts to repair, reunite, or renew a community.

Geneticists play an increasingly central role in the evolution of reconciliation projects. Sharing their expertise with communities seeking social change, they act as "biocultural brokers" who bridge scientific and sociopolitical spheres. One such individual is the renowned medical geneticist Mary-Claire King, whose activism dates to the mid-1960s when, as a doctoral student at the University of California at Berkeley, she protested against the Vietnam War and the US invasion of Cambodia. Over the next three decades, her passion for social justice and work with Las Abuelas de Plaza de Mayo (the Grandmothers of the Plaza de Mayo) would make her a pioneer in the humanitarian use of genetic analysis. She was simultaneously engaged in the pathbreaking laboratory research that would lead her to uncover the first "cancer gene," BRCA 1 (and later also BRCA 2). King became a role model for a generation of ambitious geneticists with an activist bent, including Rick Kittles, whose innovation of similar techniques would enable people of African descent to gain knowledge of their continental origins.

LAS ABUELAS DE PLAZA DE MAYO

In August 2014, Estela Barnes de Carlotto received the long-awaited news: DNA analysis had confirmed that a boy known as "Grandchild 114" had been located and was indeed the son of her late daughter, Laura Carlotto, and Walmir Oscar Montoya. Laura and Walmir were left-leaning political activists who, like scores of others in Argentina, had found themselves in the crosshairs of the military dictatorship that ruled the nation from 1976 until 1983.

In 1977 Laura was abducted by government agents and placed in a state detention center. She was pregnant at the time, and gave birth to a son the following year. Shortly thereafter she was killed by her captors. In a rare act for a regime that induced fear through "disappearance," the government returned Laura's body—which showed evidence of bullet wounds as well as disfigurement to the head and stomach, the latter an attempt to hide evidence of the pregnancy—to her mother in 1978. In 1979 the tragic loss of her daughter and her frustrated search for her grandchild inspired Carlotto to join the human rights group Las Abuelas de Plaza de Mayo, which had been founded two years prior.

When she joined Las Abuelas, Carlotto became part of a campaign that would result in the landmark application of emergent genetic technologies to a novel kind of political action: reconciliation projects. During its regime, the country's right-wing authoritarian government waged a "Dirty War" against its opponents—students, trade unionists, and others—conducted through imprisonment, torture, and murder. Estimates of the number of *desaparecidos* or "disappeared" range up to thirty thousand people. The young children of dissenters were abducted with their parents and then placed with members of the military regime or its allies. In addition, pregnant women imprisoned by the state were held hostage at detention facilities repurposed as crude maternity wards until they gave birth to their children. Afterwards, the mothers faced persecution and death, while their children were given over to allies of leaders of the authoritarian state. These children, estimated by Las Abuelas to be upwards of five hundred in number, would become known as the "misappropriated babies of the Dirty War."

Beginning in 1977 Las Abuelas gathered weekly on Thursdays in front of a government building demanding that the bodies of their children (abducted, missing, and likely murdered) and their living grandchildren be returned to them. Several members of this grassroots women's and human rights group (it later became an NGO, or nongovernmental organization) investigated the techniques that might be used to identify their families. Among the strategies they weighed was the use of then-incipient genetic analysis. In 1984, two members of the group traveled to the United States to meet with representatives from the American Association for the Advancement of Science. The activists were referred to the prominent Stanford University population geneticist Luigi Luca Cavalli-Sforza. Cavalli-Sforza in turn referred Las Abuelas to Mary-Claire King, a colleague at the University of Washington, who had spent time in Chile working as a science educator before political violence in that country made this work impossible.

Born in 1946 in a suburb of Chicago, King showed interest and talent in math and science throughout her childhood, and by the age of nineteen she had completed her BA in mathematics at Carleton College. At twenty-seven, she earned a PhD in genetics—after switching over from doctoral studies in statistics—at Berkeley. But not before she put in time fighting against social and economic inequality in the Bay Area, work that included conducting land-use research with noted consumer advocate Ralph Nader. After completing her dissertation, King served on the faculty at Berkeley for more than a decade before taking up her current post at the University of Washington in 1995. From this period to the present, extraordinary milestones in research and advocacy marked King's career. Known for her scientific philanthropy, King won renown as the geneticist who identified the genetic markers that highly predispose some to breast cancer; later she would be heralded for challenging corporations that wanted to charge large sums for the life-saving breast-cancer screening tests that were developed drawing partly on her research.[2]

What members of Las Abuelas needed was evidence that they were related to the children they claimed as family; their desire was to be reunited with their missing grandchildren who were a bridge between the activists and their lost sons and daughters. What had been rended was the biological family, and this was what the activists sought to restore even as they themselves became a chosen family wedded by heartache.

Las Abuelas required specific and indisputable proof that a grandmother and supposed grandchild were indeed related. King, aware that identification of human leukocyte antigens (HLA) was useful for matching patients needing organ transplants with compatible donors (because these proteins "help the immune system distinguish 'self' from 'non-self'") applied this analysis to Las Abuelas' mission.[3] Comparison of the HLA of the grandmothers and their suspected kin could be achieved through a simple blood test. In 1984, in the first successful case in which King was involved, a girl's HLA was closely matched—with 99.8 percent accuracy—to that of her paternal grandfather. This evidence compelled Argentina's Supreme Court to demand the return of the girl to her birth family. King and Las Abuelas had successfully repurposed a technology used for matching organ donors and recipients for a project of familial reunion and sociopolitical recovery. But while people sharing the same HLA sequence *may* be related, they are not *necessarily* related. Could a more precise method be found?

King soon began using a new technique that had been developed by her graduate school advisor and collaborator Allan Wilson.[4] With Rebecca Cann and Mark Stoneking, Wilson coauthored the trailblazing 1987 study in *Nature* that confirmed the existence of "Mitochondrial Eve"—the woman hypothesized to be the common maternal ancestor of modern humans, the genetic mother of us all.[5] Mitochondrial DNA (mtDNA), the energy catalyst of cells, is inherited by male and female children exclusively from their mothers in an unbroken genealogical line. This quality of mtDNA was proved especially fruitful for Las Abuelas' work because *all* maternal relatives, be they male or female, have the same mitochondrial sequence. This meant that biological grandmothers could be matched to biological grandchildren in the absence of the remains of "disappeared" parents and with more precision than the HLA method.[6] King thus brought the latest advances in molecular biology into the realm of contemporary political advocacy. Las Abuelas found in King a geneticist without borders, who was willing to think beyond the boundaries of the scientific lab and put DNA analysis to political and humanitarian uses that were previously unimaginable.

Begun as a small-scale project of research philanthropy, Las Abuelas' mission was transformed by the Argentinean state into a project of national reconciliation. The Argentine Forensic Anthropology Team, an NGO, was formed with the assistance of US forensic scientists in 1986 for the express purpose of locating and identifying the remains of victims of the junta state and returning them to their families. The administration of Raúl Alfonsin, the first democratically elected president of Argentina after the fall of the military dictatorship in 1983, immediately put in place the National Commission on the Disappearance of Persons. This commission produced a report the next year that detailed the brutality of the military regime, including documenting the disappearance of thousands of Argentinean citizens. In 1987, the state also established the National Genetic Data Bank, a repository of genetic samples from family members of disappeared persons that can be used for matching and identification.[7] By 2009, Argentina's Congress passed a law compelling persons to undergo DNA testing if other evidence suggested they may be a child of a disappeared person. As of 2015, more than one hundred children—now adults—had been identified and reunited with biological kin.[8] In cases in which parents had been raising children they believed to be adopted rather than stolen, shared custody arrangements were made.[9] Guido, the thirty-six-year-old grandson of Estela Carlotto, was the 114th child to be identified. Several weeks later, the granddaughter of Alicia Zubasnabar de De la Cuadra, a founding member of Las Abuelas who passed away in 2008, was discovered. She is "Grandchild 115."

Reconciliation in the case of Las Abuelas was about both kinship and the nation-state. It was about the integrity of families and the future of a country healing from trauma and political violence. With state-of-the-art methods of DNA analysis, King, working with a transnational group of scientific collaborators, helped to spearhead what are now more than three decades of initiatives in which genetic science is engaged in issues of sociopolitical or historical import. Indeed, the organization now known as La Asociación Civil Abuelas de Plaza de Mayo has been heralded as the "first group worldwide to organize around genetic technologies as tools for human rights."[10]

In the late 1990s, black researchers and activists in the United States would similarly characterize their use of genetic analysis as a human rights issue when they sought information about their African origins.

AFRICAN AMERICAN RECONCILIATION

As a boy growing up in Central Islip, on New York's Long Island, Rick Kittles demonstrated a penchant for science. As early as primary school, the Georgia-born Kittles recalls being the only black student in his classes and wondering why he looked different from his classmates. This curiosity would remain. In 1989 he earned a degree in biology from the Rochester Institute of Technology, and from there he attended George Washington University after a short stint as a high school teacher, first in Maryland and later in upstate New York. In the early 1990s, Kittles was coming into his own as a scholar-activist. While in graduate school in Washington, DC, he served as president of a "black study reading group" called Tu-Wa-Moja (which means "We Are One" in the Kiswahili language, said to be the most common on the African continent). The group did more than simply read books; they engaged in confrontational public theater, such as the day in September 1991 when they famously shut down the African Hall of the Smithsonian Museum of Natural History, believing it was racially insensitive and inaccurately reflected their history.

By the time he was a newly minted PhD, Kittles's African-centered perspective and skill as a genetic scientist had converged into a mission of reconciliation for and among black people that led to his founding of the African Ancestry company. Kittles's DNA analysis would enable the use of genetic ancestry testing as a forensic mechanism; he and his clients and collaborators took up the technical baton from Mary-Claire King and ran with it in new directions.

I have been on hand for more than a dozen presentations by Kittles over the last decade. On one such occasion, I watched in wonder as he brought a compelling combination of erudition, charisma, and folksiness to a talk before a rapt audience of over a hundred people—women and some men, mostly in their fifties and sixties. Kittles, then in his late thirties, dressed in a suit and tie and sporting a stylish goatee, delivered a lecture entitled "Trace Your DNA and Find Your Roots: The Genetic Ancestries of African Americans." In his talk, he detailed the scientific assumptions on which his company's products are based in nontechnical language accented with homey charm. Kittles considers public speaking about genetics to be part of his political work, and he feels as comfortable talking at a small community center as he does at the bench in his genetics laboratory.

Without fail, whether he is speaking to genealogists or human geneticists or a crowd made up of both, Kittles is greeted by enthusiastic audiences. A presentation on prostate cancer genetics might venture into a discussion of ancestry and how histories of social isolation and discrimination can be embodied in our DNA. A presentation touting his African Ancestry company will always include mention of the problem of racial health disparities. While some specialists might see little overlap between oncogenes and genetic ancestry, Kittles integrates them seamlessly. And the personal stakes of his research are evident in all his lectures.

Kittles's public presentations inevitably begin with a basic introduction to DNA. The "Genetics 101" section of his discussions serves to establish a pedestal of knowledge upon which his subsequent observations about the historical losses of people of African descent can be balanced. In his signature jocular style—a laid-back performance showcasing his wide-ranging erudition and humor spanning the spectrum from blue to mordant, with respect to the graver aspects of slavery—Kittles details the four chemical bases comprising the DNA molecule. He then may recount his personal frustrations with traditional genealogical research, an account followed by passionate testimony about how genetic genealogy testing had helped him discover, through analysis of his mtDNA, that he was "related" to the Hausa of Nigeria. His next statement is often similarly poignant: he will share the sobering news that his paternal Y-DNA traced to Germany. Kittles attributes this result to what he describes as "the Thomas Jefferson effect," gesturing at once to the sexual violence of slavery and to the DNA analysis that, along with archival records, strongly suggests the third US president fathered a child with Sally Hemings, a woman he enslaved.

He then describes how these As, Cs, Gs, and Ts join up in base pairs that take the spiral form of the double helix. Sometimes Kittles will note that while most DNA is located in the nucleus of cells, a small portion is found in the mitochondria and that this "maternally inherited" mtDNA is useful for understanding ancestry. On almost all occasions he ventures into the genetic paradox that human rights geneticist Mary-Claire King described to the *Los Angeles Times* as "Everybody is the same; everybody is different."[11]

Kittles explained polymorphisms this way when speaking before an audience in Chicago in 2013:

> "Poly" means "many," "morph" means "form." So, we have many different forms of the DNA. If you look out in this room, there are maybe a couple hundred people in here—there will be many different patterns in the DNA, if we compare any two people. In fact, we've cataloged over forty, fifty million of these polymorphisms. A class of polymorphism is called a "snip [SNP]," a single nucleotide polymorphism. So let's say we are looking at a track of DNA:
>
> A C T C A G T T C A

> Maybe 94 percent of you guys in the room may have a C at that second-to-last position. While about 6 percent may have a T:
>
> A C T C A G T T *T* A
>
> That's . . . a snip. Single nucleotide polymorphism. A subtle change . . . Some of them have detrimental effects in the gene, like sickle-cell disease, which is due to one polymorphism in the beta-globin gene. . . . Other diseases are due to several of these polymorphisms; they're more complex. . . . The job that I have is to try and understand how the inheritance of these polymorphisms may impact the susceptibility for disease. Now, those snips actually reflect history. They are like tags. They are like markers. And just like if you're a genealogist and you're searching in your family tree, you have these markers that reflect time periods, and who married who, and how many children they had. These polymorphisms are useful for tracing family history, too. Think about it like this: We inherit them from our parents, our parents from their parents. And we can then trace these genetic variants within a family but also within a community. And those communities within continents.[12]

Reference to "the Maafa," the compelled dispersal of African men and women from the continent of Africa, soon follows in a Kittles presentation. "Maafa," which means "great tragedy" or "great disaster" in Kiswahili—rather than "transatlantic slave trade" or "Middle Passage"—is the term preferred by many African-centered scholars and activists. Kittles's use of the term offers insight into his political commitments and to the motivations that lie behind his commercial and academic research.

In his talks focused on race, genetics, and health, Kittles may make only passing reference to the Maafa. On other occasions, however, he pauses his PowerPoint presentation—a deck of scatterplots, screenshots of scientific papers, gene flow maps, and text-heavy slides that we researchers are dissuaded from using during our talks but nevertheless do—to show the iconic image of the *Brookes* slave ship. This well-known drawing is an overhead, cross-section view of the ship, indicating the number of African men, women, and children that could be packed within its decks, basically stowed as human cargo, for the ocean crossing. The image was used by abolitionists as an indictment of the transatlantic slave trade. It stands out among Kittles's other slides; solely visual, it is intended to make this point: as much as any other application, Kittles's work has been directed toward the resolution of the Maafa. In a Kittles lecture, discussion of the Maafa counterposed with the image of the *Brookes* ship underscores his observations about the deep social damage inflicted by racial slavery.

One of the more interesting places I found myself in Kittles's company was at the Harlem temple of the Church of Latter-day Saints. Located in a historic African American community, on Malcolm X Boulevard, this Mormon church community that by doctrine formerly shunned black members now welcomes them. On this Saturday morning in February 2007, I was among a group of African American visitors interspersed with a small multicultural cohort of church members called to celebrate Black History Month through a journey of self-discovery. The words on the flyer advertising the event echoed the sentiments of a member of the Mormon church who welcomed us: "You can't know where you are headed if you don't know where you are from."

Following his presentation, audience members at the Harlem temple were offered African Ancestry's MatriClan and PatriClan tests free of charge in return for participating in a scientific research study on pigmentation. Kittles was engaged in research seeking to draw correlations between genetic characteristics and skin tone—between genotype and phenotype—with collaborators Mark Shriver, a Penn State biological anthropologist, and Charmaine Royal, a geneticist then based at Howard University. The genetic samples used for root-seeking were being repurposed for scientific research. In addition, each subject had their skin color recorded with a spectrometer. Here, years before the DNA testing company 23andMe would sell its customers' data to the pharmaceutical company Pfizer, the blurred lines of genetic testing were readily apparent. Scientific research and "recreational" DNA analysis were coming together at this event; the social life of DNA was expanding. (Several years later, when Shriver began to develop criminal forensic applications from this research, Kittles bowed out of this line of study, saying that he didn't "want to help them put more black people in jail.")[13]

There was a robust response to Kittles's research proposition and there were lines of people in every corner of the upper floor of the temple where the testing was taking place. While research assistants collected genotype and classified phenotype, Kittles and I talked in a nearby stairwell—a location that kept him close enough to the work at hand should he be needed. He shared with me the importance of his African-centered perspective on life. He stated that his most profound intellectual influences were the black studies scholars Ali Mazuri, Molefi Asante, and Cheikh Anta Diop. Indeed, Kittles would publish his first peer-reviewed paper not in a scientific publication but in the *Journal of Black Studies.* Established in 1970, and edited by the Afrocentric theorist Asante, this interdisciplinary journal is dedicated to the "dynamic, innovative, and creative research on the Black experience." Entitled "Nature, Origin, and Variation of Human Pigmentation," Kittles's article can be seen as an attempt to answer the question he pondered decades before in primary school—Why do we look different from one another?—and which he still sought to answer.

During our wide-ranging conversation, Kittles spoke of kinship structures that have been dismantled, of family ties that were cut, of broken links to the African continent. He offers his company's genetic ancestry testing as a kind of homecoming, as a way to put the pieces back together, to bind the broken limbs of family trees. On another occasion, speaking about what inspires his research and his work with African Ancestry, he stated: "How often do you hear African Americans talking about Africa in a positive light? . . . Our history doesn't start with slavery; we came through slavery, but so many of our youth don't understand that . . . [and] they are the ones who are going to shape this reconciliation that's occurring." When Kittles presented the keynote address to the United Nations General Assembly on the occasion of the International Day of Remembrance of Victims of Slavery and the Transatlantic Slave Trade in 2012, he similarly declared that he started his company because "I wanted to bring African Americans closer to Africa." For Kittles, genetic ancestry testing of black Americans may hold the key to the reconciliation of Africa and its diaspora.

TRUTH AND RECONCILIATION POLITICS

DNA analysis is today imagined as a medium through which societies may move toward truth and healing. These reconciliation projects comprise a wide spectrum of social actions and anticipated outcomes. And these projects may not have a practical or legal conclusion; indeed, their end points may be unspecified. Or, as political scientist Melissa Nobles observes of official state apologies, the "critical reexaminations of history" that may be occasioned through these practices allow us to appreciate the "ideological and moral stakes" that are being expressed and not focus solely "on anticipated material gains or losses."[14] Therefore, the inauguration of a reconciliation project may in itself warrant our attention.

The word "reconciliation" readily brings to mind the South African Truth and Reconciliation Commission and the poignant public hearings in post-apartheid South Africa that began in 1996 and continued for four years. During these hearings, victims testified about their experience with apartheid state violence. (Some perpetrators of violence confessed their offenses and—controversially—were given amnesty from prosecution.)

Less well known is one of the United States' own truth and reconciliation processes, which suggests the intensity with which racial healing is still being sought generations after the cessation of the transatlantic slave trade and decades after the civil rights movement of the mid-twentieth century. As in South Africa, testimony was the forensic vehicle. Although our interest here is in the social life of genetic ancestry testing, this US example sheds light on racial reconciliation strategies more broadly.

Inspired by the South African example, in 1999 the citizens of Greensboro, South Carolina, inaugurated a truth and reconciliation process—the first ever in the United States—in an attempt to confront and dismantle long-standing racial tensions between blacks and whites in that community. While the mechanism of truth differed between Greensboro and laboratories in Buenos Aires and Washington, DC, what was strongly shared was the sentiment that there is little hope of a tenable future for these communities if these historical injuries are not attended to and repaired. Similar to the Venture Smith endeavor, whose descendants sought societal healing, it emerged out of recognition that unresolved issues of racial discrimination would not simply dissipate over time. Notably, Greensboro is important in the history of US racial politics because it was here, on February 1, 1960, that the sit-in strategy of the modern civil rights movement was launched. Four brave African American students from North Carolina Agricultural and Technical State University struck a blow against Jim Crow by ultimately succeeding—despite violent antagonism—in integrating the lunch counter of a local Woolworth store.

The Greensboro Truth and Reconciliation Commission, however, responded to a lesser-known demonstration against racial and economic inequality almost two decades later, which ended tragically. In November 1979, a multiracial group of activists were beaten and fired upon by members of the Ku Klux Klan as they protested in a local public housing neighborhood in a legally sanctioned "Death to the Klan" march. The attack left five protestors dead and ten wounded. Local news cameras captured the assault on film. Astonishingly, police on the scene arrested one marcher, while the shooters fled the scene. After the arrest, the protestor's bail was set at double that of those accused of murder and assault. With acquittals of the alleged perpetrators despite two trials, the "Greensboro Massacre" left behind festering resentment, immense pain, and profound feelings of injustice.

The Greensboro Truth and Reconciliation Project began in 1999 on the twentieth anniversary of these events. After two decades, citizens from many facets of the community—including victims and survivors of the attack, college students, churches, NGOs, and civil rights activists—acknowledged that this incident remained an open wound in Greensboro and that work toward healing was essential. The reconciliation process stemmed from its backers' recognition that the racist culture that spurred the violence "continues to effect [*sic*] the quality of economic, social, political, spiritual and educational life in Greensboro."[15] The project's organizers contend that "confronting and reckoning with the past is necessary" if the city and the country is to move forward regarding race relations. Or as it is put in the preamble to the project's official mandate: "There comes a time in the life of every community when it must look humbly and seriously into its past in order to provide the best possible foundation for moving into a future based on healing and hope."[16]

The Greensboro Truth and Reconciliation Project was the first time the process had been carried out in the United States, and the occasion also marked the first time in the world that the process was based in a city rather than a nation-state. Despite "fear of and hostility toward" the process, several years of research, retreats, assemblies, and two hundred interviews in Greensboro culminated in a series of hearings in 2005.[17] A final report released the next year concluded that racial violence harmed the community in addition to the individuals who were murdered and assaulted and recommended that both the City of Greensboro and the perpetrators of violence formally acknowledge the events of November 3, 1979. It also recommended institutional reforms in city government, the courts, and the criminal justice system.

In Greensboro, the forensic mechanism was the multiple voicing of memories and experiences layered upon one another to come to a better understanding of harm, trauma, and discrimination, with the hope of a better future. While the South African truth and reconciliation process has been criticized as ineffective, it remains an obvious example of what has yet to be even attempted in the United States on a national scale—a public conversation about the history of racism in this country. And so, resolution of these issues is sought in other ways, including a repurposing of genetic ancestry testing techniques that finds its inspiration in Greensboro, Buenos Aires, and elsewhere.

We might think of African Ancestry's genetic kits as grassroots politics writ small, as efforts at small-scale truth and reconciliation practices in which DNA test results are used as a form of testimony. Although there are notable differences with the work that transpired in both Buenos Aires and Greensboro, the uses of genetic ancestry testing that I trace in the pages to follow emerge from similar impulses to prompt awareness of injurious histories, and their legacies today, as a necessary precondition for charting a new way forward.

Genetics has become a medium through which the unsettled past is reconciled. The pursuit of historical and social repair that undergirds reconciliation projects has been described as a kind of politics of the past that risks dwelling there.[18] Yet what is distinctive about reconciliation projects—whether the forensic mechanism is testimony or technology—is precisely the combination of historical reckoning and future orientation they effect. Future promise, anthropologist Michael Fortun tells us, is "an eradicable feature of genomics," whether with respect to our predisposition for certain diseases or our prospects for our communities.[19] Reconciliation projects are driven by the desire to effect change in the present and to shape a different future. Closure is not necessarily what is sought, for, as the moral philosopher Susan Dwyer states, reconciliation's goal is largely "to make sense of injuries" in the general narrative of a person or nation's life.[20] Rather than miring us in the past, these efforts forensically excavate it to offer the hope of new social and political possibilities.

TWO

Ground Work

> It's bad enough that some of the bodies that may be in those tombs were discriminated against in life. But now, they're being discriminated against in death.

In 1991 archaeologists uncovered several graves on a plot in lower Manhattan. These burials were discovered during completion of a land survey conducted on behalf of the US General Services Administration (GSA), which planned to construct a government office tower at the location. Initial archival research had suggested that the building's proposed location might be the site of a colonial-era cemetery. The survey was undertaken in compliance with the National Historic Preservation Act of 1966 (NHPA), which mandates protection of historic properties and burial remains at proposed building sites utilizing federal funds. The survey confirmed that the location was the former site of the "Negros Buriel Ground," a municipal cemetery for the city's African and African American population that dated to the colonial period.

This site is now the African Burial Ground National Monument, administered by the US National Park Service. Its status as a national monument reflects the federal government's belief that it has special "historic or scientific interest."[1] Yet this burial ground holds further significance still. Its disposition proved foundational to direct-to-consumer genetics in the United States.

Commercial genetic analysis is aptly regarded as an offshoot of the Human Genome Project, completed in 2003. However, an origin story for genetic testing that begins with single nucleotide polymorphisms ("SNPs") or supercomputing can only partly account for why this analysis became important in African American cultural politics. As we will see, the controversy that transpired over excavation methods and research priorities at the centuries-old African Burial Ground shaped subsequent reconciliation projects incorporating DNA that aimed to address the history of chattel slavery and, in so doing, reframe the racial politics of the moment.

The African Burial Ground initiative illustrates how the social life of DNA is forged both technically and conceptually. Although the remains excavated at the site were analyzed using several methodologies, most notable was a then-novel use of genetic analysis. Second, research undertaken at the site would become paradigmatic for how genetics could be used to create new identity and reconstruct the past. Although studies of ancient genes (aDNA) had been conducted in the late 1980s—beginning with research outlined in *Nature* by geneticist Bryan Sykes and colleagues demonstrating the ability to amplify DNA obtained from bone—the African Burial Ground project was among the first and most public uses of aDNA analysis in the United States.[2] Here, genetic comparison was used to infer the ancestral associations and ethnic affiliations of the individuals buried at the site. Third, the undertaking was the inspiration for subsequent commercial endeavors. The African Burial Ground project led to the formation of African Ancestry, one of the earliest genetic-ancestry-testing ventures, when Rick Kittles, then a geneticist working on the project, converted the research techniques utilized in the enterprise into a successful business.

UNCOVERING THE "NEGROS BURIEL GROUND"

The graves unearthed in Manhattan in February 1991 were part of a municipal cemetery for the city's African and African American population dating to at least the late 1600s. This segregated cemetery, once known as the "Negros Buriel Ground," had been established for blacks in New Amsterdam (later New York), who were not permitted to bury their dead within the city's walls by Dutch (and later English) colonial authorities. Under a policy that has retrospectively been characterized as "mortuary apartheid," throughout the seventeenth and eighteenth centuries, the burials of free and bonded Africans were relegated to this isolated ravine located outside municipal boundaries.[3] By the nineteenth century, the burial ground and the surrounding area had been incorporated into the city limits and covered with landfill in order to make way for the expansion and development of lower Manhattan—an area that now includes New York's City Hall, the Wall Street financial district, and the exclusive Tribeca neighborhood.

While both government officials and the New Jersey–based archaeological salvage company Historic Conservation and Interpretation—the company conducting the land survey—were aware of the presence of the graveyard, the uncovering of hundreds of intact burials at the site was nevertheless surprising because archaeologists hypothesized that most remains would have been destroyed long ago. As a Historic Conservation report explained, the company's archaeologists believed that subsequent construction and development on and around the site in the years since its closure in the 1790s had "obliterated any remains . . . within the historic bounds of the cemetery."[4] Accordingly, the conservation company foresaw no regulatory hurdles to the smooth start of the GSA's construction process.

Historic Conservation's pre-construction survey of the cemetery was conducted using the so-called "coroner's method."[5] With this type of removal, the graves were unearthed using large construction machinery, with little consideration given to conservation of the remains, the composition of the burial, or the material culture included in the gravesites. Having been gathered in the pell-mell fashion characteristic of the coroner's approach, the contents of the graves were then transported to the laboratory of the Metropolitan Forensic Anthropology Team (MFAT) based at Lehman College, a campus of the City University of New York. To make matters worse, after reports of Historic Conservation's rushed excavation came to light, the company's archaeologists, hired to discern the salvageability of the African Burial Ground, were accused of contributing to its destruction.[6]

News of the coarse treatment of the burials angered the local community. Activists, politicians, and preservationists were dismayed that they were not immediately informed of the discovery of the cemetery and only learned of its existence through the media. Community groups and cultural organizations, including one that called itself Descendants of the African Burial Ground, were particularly outraged by what they saw as the state-sanctioned disruption of sacred ground. Some attributed the poor handling of the graveyard to racial prejudice. State Senator David A. Paterson (later governor of New York), who organized the Task Force for the Oversight of the African Burial Ground, for example, protested: "It's bad enough that some of the bodies that may be in those tombs were discriminated against in life. But now, they're being discriminated against in death."[7] Numerous efforts sprang up to bring attention to these grievances and correct them, ranging from mainstream politicking and protests to the enactment of African funerary rites at the burial-ground site to settle disturbed souls.

Though they were of varied ideological orientations, the burial ground's stakeholders were mostly united in their intention to intervene in the process to determine the future of the gravesite. To do so, they fastened upon a provision of the National Historic Preservation Act that prescribed local communities' consultation with respect to historic locations. Activists, including Ayo Harrington, Elombe Brath, Miriam Francis, and Earl Maitland, demanded (and received) a role in the excavation and administration of the graveyard.[8]

In response to public outcry, the excavation of the African Burial Ground proceeded using the more deliberate and time-intensive methods of scholarly archaeological practice—as opposed to forensic or commercial archaeology. With the former, the remains were "intricately measured and delicately removed from the land," in contrast to the rushed pace of the coroner's method.[9] Changes in technique and in the decision-making process, however, did not fully diminish the bad faith that had been established between the GSA and the local community. Reacting to reports of the new mode of excavation, Senator Paterson conspiratorially opined to the press that while he was heartened that the work would now be more carefully executed, he was "not assured that this was the original intention" of the planners.[10]

Neither did the change of course prevent continued damage to the remains: despite new procedural safeguards, on February 14, 1992, an employee working for New York City's Landmarks Preservation Commission witnessed human bones being jostled from gravesites, and other remains lying in the bucket of a backhoe. Furthermore, despite the employ of more disciplined excavation methods, several burials were destroyed when a construction worker accidentally poured concrete on the graves. There were also reports that remains were being improperly conserved. At Lehman College's lab, bones from the burials were allegedly wrapped in newspaper rather than in acid-free, conservation-standard materials. These remains were reportedly also kept "under improper environmental conditions" and were "inadequately stored on top of each other."[11]

The haphazard handling of the burials prompted Mayor David Dinkins—the first black American to hold New York City's highest office—to confront federal authorities. The mayor's charge that the GSA had violated the National Historic Preservation Act—as well as its Memorandum of Agreement with the city and its historic and landmark preservation agencies—spurred national congressional hearings. Congressman Gus Savage, a Democrat from Illinois, who was sympathetic to the perspective of Mayor Dinkins and the activist community, helmed a subcommittee that ruled that the GSA (and, by association, Historic Conservation and Interpretation) had violated the NHPA by not developing an adequate scientific research plan for the site and ordered that work there cease. As a consequence of activist foment and political pressure on both local and national levels, excavation at the African Burial Ground was permanently stopped in July 1992.[12] By this time, remains from more than 419 burials (of 15,000 to 20,000 estimated total burials at the cemetery) had been transported to the MFAT laboratory at Lehman College.[13]

Emboldened by their victory in ending the excavation process, some stakeholders now made additional demands. They insisted on being consulted about the planned study of the burials, and supported research that could yield information about the history of the individuals buried there, and about the communities they may have originated from before being transported to the Americas for enslavement. It was hoped that such a research agenda could produce information about the national or ethnic origins of the buried persons (and, by association, those of some black Americans). Additionally, the activists lobbied for Michael Blakey, a leading physical anthropologist and an African American, to take over efforts to study and conserve the African Burial Ground remains. Blakey in turn encouraged the activists' stewardship of the site.

It is notable, however, that some members of the local community were not solely or even primarily interested in the African Burial Ground's historical and scientific potential. A significant faction saw the site first and foremost as a cemetery, and therefore a sacred site. Glen C. Campbell, an African American architect who consulted on the research design for the African Burial Ground study, stressed this point in a letter to the GSA. He wrote, "The human remains and the associated artifacts are severely important, *but the place is sacred.* No one would dare suggest digging ten or twelve feet below grade at the Arlington National Cemetery and moving the remains. . . . To bury the remains elsewhere will be nothing short of disrespect for the community and the sacredness of their ancestors' contributions."[14] In recognition of the site's spiritual significance, activists initiated a vigil of drumming, chanting, and prayer.

In October 1992, Blakey was named scientific director of the research project, and early the next year the excavated African Burial Ground remains were transferred from the MFAT in New York City to the W. Montague Cobb Research Laboratory at Howard University, a private, historically black college in Washington, DC. The remains would be stored and studied in accordance with a research design plan drafted by Blakey, approved by both government agencies and the local concerned community, and funded with $6 million from the GSA.[15] The historical and social merit of the African Burial Ground was certified by its designation as a National Historic Landmark that same year. Among researchers, however, conflicting interpretations of the site's historical and social value would remain.

THE CASE AGAINST "BIOLOGICAL RACING"

At the Lehman lab, the method of analysis consisted primarily of osteology—the scientific measurement of the skeletal remains—and the broad classification of them into several categories, including stature, age, sex, and race.[16] Male and female bones largely cluster into distinct weight and length classes. Moreover, some forensic osteologists contend that blacks have longer femurs than whites and Asians and, therefore, bone length has been used for black versus nonblack racial classification.[17] This forensic approach was and remains standard practice among many physical anthropologists and was the perspective that the MFAT scientists brought to the research project.[18]

Other physical anthropologists, however, including those at Howard University who would become involved in the African Burial Ground project, found the forensic mode of analysis and interpretation inadequate to the historical significance of the cemetery. Detractors of the Lehman approach, including Blakey, the burial ground's new research director, contended that these studies were unduly preoccupied with the gross racial classification of the sort employed for criminal justice purposes.[19] He further maintained that this methodology reduced the individuals in the burials to "narrow typologies" and thinly "descriptive variables," and thereby "disassociated" them from their "particular culture or history."[20] In an article with Cheryl J. LaRoche, a conservation consultant for the US National Park Service, Blakey asserted that the Lehman scientists' forensic approach—which focused on gross classification—severely "underestimated the enormous analytic value of the cemetery site."[21]

Howard researchers, by contrast, shared the conviction that the cemetery was an exceptional discovery. The Howard team's research proposal noted that the site was "the earliest excavated municipal cemetery in the United States" and "the largest African American archaeological population currently known." Blakey would expound on the site's importance in his concluding report on the African Burial Ground project, stating, "The skeletons, artifacts, and documents of the people buried [in the cemetery] tell volumes about their lives."[22] If questions about the burials were aptly framed, Blakey and his colleagues believed, details about the preparation or positioning of bodies in the burials and the placement of funerary ritual objects found at the site, combined with scientific analysis of the remains, could open a rare window on the experiences of blacks in the Americas and perhaps shed light on their ethnic origins as well.

This disagreement over research methodologies reflected ongoing debates in the discipline of anthropology over whether "descriptive" or "analytical" studies of skeletal remains were most useful for reconstructing the past. Descriptive (or forensic) studies are those that primarily involve "sorting" and "identification" without concern for the "broader theoretical context."[23] Today, anthropologists rely on both analytical and descriptive approaches.

In New York, community members took up their own positions in the descriptive versus analytical debate. Activists associated with the Descendants of the African Burial Ground complained that the approach taken by researchers at the MFAT lab amounted to the "biological racing" of their ancestors' bodies;[24] they expressed vehement opposition to classification that they believed would "reduce their ancestors' social identity to skin color."[25] The lens through which the burials would be interpreted—not to mention the cultural background of the scientists doing the interpreting—had been a key motivator for activists calling for the transfer of the remains from MFAT to Howard. Blakey engaged MFAT lab director James V. Taylor directly on this point, writing to him in December of 1992 to express his dissatisfaction with what he deemed the Lehman-affiliated researchers' unnecessary reliance on racial categories. Notably, this exchange was also an opportunity for Blakey to express why he felt the MFAT's demands for further involvement in the research should be denied. Responding to Taylor's prior correspondence, Blakey wrote,

> Your contention that race estimation is essential to determining age and sex is usually supported by researchers of a previous era in the history of anthropology. . . . "Races," such as "negroids, caucasoids, and mongoloids," which you wish to apply have come to be understood as folk taxonomy, social constructs imposed on a natural world whose genetic variation is far more complex. . . . The study of specific population affinities is more important and accurate than gross racial classification.[26]

A compromise was proposed in which the MFAT would be allowed to complete its work, following which the remains would be turned over to the Howard team for analytical and interpretive analysis.

Criticism of the "biological racing" of the remains also suggested awareness on the part of both activists and scholars of the historical use of biometrics—the measurement and analysis of human characteristics as a means of grouping individuals—to bolster scientific racism, and thus the potential for flatly descriptive work to yield prejudiced interpretations of the burials. For, as Stephen Jay Gould and numerous others have documented, the comparative "mismeasurement" of bodies, from lung capacity to crania to genes—with white bodies serving as the norm against which all others are measured—has long been employed to advance erroneous claims about black inferiority.[27]

Against the backdrop of this bitter legacy of discriminatory biological research, supporters of the Howard team's analytical method of interpretation sought to upend this history by using biometrics, alongside other forms of scientific and humanistic analysis, to glean new information about the physical effects of slavery as well as the African origins of some of the earliest black Americans. Indeed, for the self-described "descendent community," a name suggested by the Howard researchers, the discovery of the African Burial Ground represented a stirring possibility—a retreat from designation by "skin color" alone and the stigmatizing concept of race, for both their "ancestors" and themselves.

For some supporters of the analytical approach to the African Burial Ground remains, the stakes were very high, for this also presented an opportunity to restore ethnic identities to a racialized (and racially subjugated) community. Historian Michael Gomez has shown that the Africans brought to the Americas via the tortuous journey known as the Middle Passage "exchanged their country marks" in a process of compelled racialization. This process, Gomez writes, was a transition from an "ethnically based identity directly tied to a specific land to an identity predicated on the concept of race."[28] Although some enslaved Africans would strive to retain distinct practices and identities that reflected the perseverance of their "country marks," they nevertheless were compelled by their states of bondage "to learn the significance of race," and to cope with their racialized caste position in slave societies such as the United States.[29] In walking a fine line in which they contested the "biological racing" of the remains, yet advocated for the use of similar scientific techniques toward the construction of fuller interpretations of the slave past, the activists and Howard researchers engaged in a quest for the reversal of the racialization produced by slavery—if not its enduring effects.

More specifically, the politics on the ground of this Manhattan cemetery split hairs between race and ethnicity. The activists sought to restore pre-enslavement identity to the individuals interred at the burial ground. This restoration was of benefit to the descendant community and potentially to the body politic as well. The distinction between descriptive and analytical approaches to interpretation, between "biological racing" and the restoration of specific details of African origins that we might describe as ethnicity, was the backdrop for the employ of scientific analysis that would in a few years' time be born out in genetic ancestry testing.

The eventual siting of the African Burial Ground research at Howard University marked a fundamental change in the framing of how and why the research was conducted. The question undergirding the investigations carried out at the MFAT lab could be summarized as "Are these the bones of blacks?"[30] The Howard researchers, to the contrary, sought answers to a more extensive set of questions, including "What are the origins of the population, what was their physical quality of life, and what can the site reveal about the biological and cultural transition from African to African-American identities?"[31] In posing these questions about the remains, Blakey's team hoped to use these rare remnants of black colonial life as an opportunity to more fully detail knowledge about how those buried at the African cemetery in lower Manhattan lived and died.[32] At the Howard lab, in other words, the research orientation was shifted from an epistemology of racial classification to an epistemology of ethnicity (and therefore, also ancestry).[33] Indeed, the African Burial Ground researchers' write-up of the study noted their interest in discerning the "rainbow of [African] ethnicities" that might be found at the cemetery. Analysis of the remains was thus broadened to include a panoply of social and historical interpretation that might render "biological evidence of [the] geographical and macroethnic affiliations" of enslaved Africans in colonial New York.[34] Howard University's Cobb laboratory, a research site with a long history of comparative anthropological and archaeological study of African and African American remains, was a fitting location for the African Burial Ground project. It was also the case that this project promised to bring both funding and renown to the Washington, DC–based institution—and to its researchers.

THREE

Game Changer

> The scientific research now underway constitutes yet another dimension of a long-standing human rights struggle among African Americans. . . . We seek to restore knowledge of . . . African-American origins and identities.

One of a handful of African American biological anthropologists, Michael Blakey has a passion for the material culture of the past that was ignited in his childhood. His biologist mother had African American and Native American ancestry; she was related to the Nanticoke Moors of Delaware, a mixed-race community dating to the 1700s. In his youth, Blakey spent a good deal of time in Delaware with a great uncle who "walk[ed] the fields and collect[ed] artifacts from their ancestors. . . . He was one of the biggest pothunters in Delaware. This was one of my favorite things to do. . . . My hero was [anthropologist] Louis S. B. Leakey. . . . I was in all-black schools and public schools and it was a strange thing [to be interested in]. But that's what I was about."[1] By the age of fifteen, he was working as an archaeological intern at the Smithsonian Institution.

After attending college at Howard University, where he majored in anthropology with an emphasis on Africana and Mesoamerican studies, Blakey earned a master's degree and a doctoral degree in anthropology at the University of Massachusetts at Amherst, in 1980 and 1985 respectively, gaining experience in bioarchaeology and physical anthropology during his course of study. There his dissertation focused on "the political economy of psycho-physiological stress" on living persons in the United States and the United Kingdom in order to better "understand how racism and class affect stresses that lead to things like elevated hypertension rates."[2] The author of scores of scholarly publications, Blakey also coedited an influential volume on "the socio-politics of archaeology" in 1983. A man of wide-ranging research interests and talents, Blakey was involved with research on remains excavated from a cemetery at Philadelphia's First African Baptist Church in the 1980s, which, until the discovery of the New York African Burial Ground, was the largest US archaeological site of African and African American remains. The experiences would prove germane to his work at the African Burial Ground.

Blakey submitted his research proposal to Howard Dodson, director of the Schomburg Center for Research in Black Culture at the New York Public Library, who was serving as the chairman of Mayor David Dinkins's Advisory Council on the African Burial Ground. After the proposal was approved, the Howard researchers embarked upon an ambitious interdisciplinary investigation of the African Burial Ground remains proceeding from Blakey's June 1992 project design.

In studying the African Burial Ground remains, Blakey's research unit combined methods and insights from the social and biological sciences to generate "historically and ethnographically informed interpretations" of the origins and life course of the persons laid to rest at the burial ground more than two hundred years prior.[3] At bottom, the intent of these efforts, for both scientists and activists, was to explore these fundamental questions: What could we know about these 419 African-descended people buried in lower Manhattan over two hundred years ago? How and why did they die? From where in Africa might they have hailed? In order to shed light on the experiences and identities of the deceased men, women, and children, the GSA funded an interdisciplinary research project, based at Howard and staffed by specialists drawn from the fields of archaeology, biology, biological anthropology, cultural anthropology, and history.[4] Among the methods employed by Howard's researchers in studying the centuries-old remains were craniometric analysis, dental morphology, and molecular genetic assessment—a then-emerging form of DNA analysis.[5] Researchers hoped that this multidisciplinary approach would redirect the research mission from the narrowly forensic trajectory of the Lehman College team in order to return to today's slave descendants the "country marks" of their symbolic ancestors, which had been obliterated first during their lives and then again in their burial.

CRANIOMETRIC ANALYSIS

The Blakey team engaged in skeletal analysis of the remains recovered from the African Burial Ground site to shed light on the experiences of these individuals—insights into nutrition, disease, and the physical toll of enslavement. As part of this inquiry, researchers conducted analysis of twenty-seven recovered skulls, using "morphometric" techniques including the measurement of crania—a method for which the Lehman researchers had come under fire.[6] However, the dramatic shift of interpretive register from race to ethnicity and ancestry that accompanied the project's move to Howard was evident in the Howard researchers' utilization of these methods. What was at stake was not the techniques but the interpretative filter. For Blakey and his team, these forensic tools were useful if the data were appropriately interpreted—that is, analyzed with an eye for uncovering richer details than mere racial typing would allow.

This shift of register is similar to the kinds of strategic, judicious scientific engagement that historians Nancy Leys Stepan and Sander L. Gilman refer to as "recontextualization." Given the tragic history of scientific and medical abuses against minority groups, including the surgical experimentation on enslaved men and women in the nineteenth century and the horrors of Nazi medicine in the early twentieth century, biomedical researchers of African and Jewish descent developed strategies for challenging scientific racism while sustaining a practice of scientific inquiry. With recontextualization, the "tools of science were used either to prove that the supposed factual data upon which the stereotypes of racial inferiority were based were wrong, or to generate 'new' facts on which different claims could be made."[7] In the African Burial Ground research project, Blakey and his team applied both tactics. Given activists' and researchers' concerns over "biological racing" of the remains, recontextualization helps us to understand how the Howard team were able to be both scientific skeptics and scientists. While the investigations involved the measure of skulls, these measurements were done alongside "extensive" collaboration "with project historians and archaeologists" to produce "biocultural, interdisciplinary" data.[8]

For example, researchers observed that the heads of the burial coffins faced west (perhaps to follow the direction of the setting sun). One coffin was embellished with a *sankofa*, a West African heart-like image that symbolizes the importance of knowing about one's past in order to move ahead to the future. The bodies were wrapped in shrouding, a practice that may have indicated the buried individual was a Muslim. Many of the burials contained cowrie shells, the currency of sub-Saharan Africa, as well as glass beads from this region. Bodies were also laid to rest with coins, buttons, and jewelry. This and other archaeological evidence strongly supported the conclusion that the individuals hailed from the African continent.[9]

The team led by Blakey did not use "race" as a category of analysis. Rather they employed a broad range of analytical groupings. Carried out by team researchers Alain Froment (a biological anthropologist), Shomarka Omar Yahya Keita (a physical and cultural anthropologist), and Kenya Shujaa (an osteologist and lab technician), this mode of analysis "quantified craniometric diversity" in the burial ground sample and then compared this data to "a broad range of historical and modern African and non-African groups." As the researchers detailed in their 2004 report to the GSA, with their craniometric studies, "no reference to any 'racial' definition was made." Analysis was employed that showed similarities and differences between buried individuals based on volume, weight, and topography "without use of closed biological categories."[10]

DENTAL MORPHOLOGY

The presence of intact skulls in several graves permitted the Howard researchers to conduct dental morphology studies, relying on the examination of chemical traces in the teeth; "chemical signatures" can reveal information about a deceased individual's native origins or environment. The Howard researchers had examined the skulls removed from the African Burial Ground for strontium, a chemical isotope that varies by geological context. Likewise, biological anthropologist Alan

Goodman and his colleagues examined the teeth for chemical traces that might suggest "where in the world individuals' childhoods were spent."[11] The presence in the bones and teeth of strontium 86 and strontium 87 indicated that the deceased were born and reared in West Africa before being brought to the United States. In a few other instances, the presence of different strontium isotopes suggested nativity in the Americas. Thus, in keeping with the goals of the research project under Blakey, the dental morphology analysis was used to shed light on the particular origins of those buried in the African cemetery. The findings offered scientific proof of ethnicity rooted in West Africa, rather than racial identity alone.

In addition to information gleaned from chemical traces, physical dental traits (including grooves, cusps, enamel, and roots) can also suggest distinctive human groups and therefore help to ascertain the origins of some of the individuals whose graves were found at the African Burial Ground. Moreover, the shared presence of dental characteristics could suggest familial relationships among the several hundred burial remains uncovered. The researchers also analyzed "styles of dental modification," including chipping and aesthetic filing, known from archaeological and ethnographic research to be characteristic of certain African regions and cultures. The analysis of the chemistry of dental enamel aided as well in gaining insight into health conditions including anemia and malnutrition. With dental analysis, researchers were able to garner information about the general health of the individuals while making gains toward inferring the cultural practices and regional origins of the deceased.[12]

MOLECULAR GENETIC ASSESSMENT

Perhaps most significantly, at the Howard lab, the social and biological scientists analyzing the African Burial Ground remains made use of relatively new methods of "molecular genetic assessment" to uncover the possible origins of the cemetery population. The researchers proposed that DNA studies could aid in the determination of "genetic affinities" between the individual burials and "specific cultural/regional origins in Africa."[13] Examinations of mtDNA and Y-DNA sequences would be examined to elicit haplogroup (clusters of gene sequence variants that are inherited together) designations that might suggest "geographically distinct lineages."[14]

Blakey and his collaborators also hoped that analysis of the remains might yield some information about the health profiles of the persons buried at the site. How did the conditions of chattel slavery affect the health of those buried and how might genetic data drawn from this population shed light on black health today? The aim was to "establish the 'baseline' biology of the African American population in the United States" in order to better understand these questions.[15]

Although Blakey's team opposed "biological racing," they were open to physiological and genetic research that might shed light on clinal differences between social groups—that is, a perspective that considered human variation across a broad spectrum. By employing such an approach, the Howard researchers held out the hope that there was "potential for determining the ancestries of the African Burial Ground populations" without hewing to problematic racial essentialism.[16] Fatimah Jackson, a coordinator on the Howard team, who would later become the study's associate director for genetics, expressed hope that this research would yield the "likely ancestral homeland regions" of the deceased individuals at the burial ground, by means of a "thorough evaluation of extended haplotypes" in DNA extracted from bone.[17]

In these years when the mapping of the human genome, a process that used tissue from living individuals, was still plodding toward completion, there was little confidence in some quarters that the African Burial Ground researchers could accomplish the feat of analyzing the DNA of long-deceased persons. Not everyone was convinced that the Howard team's molecular genetic assessment plan was a viable one. A reviewer of the Howard research proposal carped that the "feasibility and efficacy" of this assessment was "questionable." This critic went on to contend that there are

> few populations available in Africa to study for comparative purposes. . . . While it will probably soon be possible to extract usable DNA from ancient bones, there have been no truly convincing demonstrations of such extraction to date. A major problem in the extraction of ancient DNA is that of excavator contamination. It is

> a safe assumption that all the exposed bones from the African Burial Ground have been contaminated with excavator DNA, despite precautions taken during the fieldwork.[18]

A GSA peer-review panel similarly concluded that exposure to the elements that caused degradation and contamination of any remaining DNA threatened the viability of the genetic analysis.[19] As the Howard researchers would discover, it would indeed prove challenging to extract, amplify, sequence, and analyze what genetic residue remained in the burials.

Howard University geneticist Matthew George carried out an initial feasibility study of fifteen of the burials in 1995. Trained at the University of California at Berkeley, where he took his PhD in 1982, George was an associate professor of biochemistry in the College of Medicine at Howard. His early research "contributed to the developments leading to the . . . hypothesis for the African origins of the earliest humans."[20] Indeed, George was a coauthor of an influential 1985 "mitochondrial Eve" paper, lead-authored by Mary-Claire King's advisor, Allan Wilson. Like King, George was trained by Wilson and did his doctoral work in this senior scientist's lab.[21]

George's efforts met with limited success. While the extraction of DNA from skeletal or fossil remains has become widely possible, it still remains a delicate endeavor. In the early 1990s this was all the more true: the University of Oxford's Alan Cooper and the Max Planck Institute's Hendrik Poinar wrote in *Science* that "ancient DNA research presents extreme technical difficulties because of the minute amounts and degraded nature of surviving DNA and the exceptional risk of contamination."[22]

Efforts to analyze ancient DNA had first been conducted in the late 1980s by geneticist Bryan Sykes and his colleagues in an important study demonstrating the ability to amplify DNA obtained from bone. Ancient DNA research relies upon analysis of mtDNA because it is very stable, and under ideal environmental conditions remains as viable biological evidence for thousands of years. Of fifteen attempts, George and his team were able to derive aDNA from "nine 200-year-old hair and bone samples."[23] Of this group, researchers succeeded in isolating the mtDNA of just four individuals. However, they were not able to clone these sequences for further study nor did they have a population of reference samples with which to compare the four.

In 1998 Rick Kittles, who was not an original member of the African Burial Ground research team but participated in the study from 1995 to 1999, while he was a graduate student at George Washington University, embarked on a new wave of genetic analysis that would prove much more successful than that of a few years prior, and in time make important contributions to the use of DNA analysis to infer the ethnicity and origins of the burials.[24] As it was put in the final New York African Burial Ground research report, "Kittles was able to bring an updated methodology to the project."[25] Kittles's skills proved to be a game changer for this facet of the research project.

In order to examine the genetic data, Kittles needed to solve two problems: First, ancient DNA sequences are fragile and short in length, and Kittles would need to find a way to manipulate the sequences that were available to him without destroying them. And, second, in order to interpret these sequences with a measure of accuracy, he would need a comparative reference database.[26]

To solve the first problem, Kittles turned to technologies that allowed for the analysis of characteristic genetic markers in Y-DNA and mtDNA. Y-DNA is unique for its consistency, passing virtually unchanged from fathers to sons, and can therefore be used to trace a direct line of male ancestors. Following the methods of researchers in human genetics such as Mark Jobling and Chris Tyler-Smith, Kittles aimed to show how genetic polymorphisms (different forms of DNA or RNA produced as a result of substitutions or deletions) in the Y chromosome could be used to trace paternal ancestry intergenerationally and demonstrate that these small changes could be characteristic of distinct social groups.[27] Even non-coding regions of DNA—called "junk DNA" because they have no known function—proved informative. Within these non-coding areas, short segments of DNA are duplicated in a pattern called short tandem repeat (STR). The number and specificity of STR markers can be used to distinguish between individuals as well as to discern ancestral relationships, because a male's Y-DNA STR is shared by individuals in his paternal line; that is, men who descend from the same male ancestor will have similar STR patterns.

Mitochondrial DNA is transferred mostly unchanged from mothers to children, so this, the smallest of all human chromosomes, can be used to uncover matrilineage. If mtDNA is preserved under ideal conditions—protected from air, moisture, and other corrosive elements—it can be extracted centuries later. It is this quality of mtDNA that has been of crucial use in instances in which ancient remains are under investigation, from long-dead flora and fauna to the deceased members of the Romanov clan.

Using a method that has since become commonplace, Kittles compared DNA from a second sample of African Burial Ground remains with that of contemporary Africans. If the sample and the reference DNA matched at a set number or sequence of genetic markers, this individual was said to have shared a distant maternal or paternal ancestor with the person who was the source of the matching sample in the reference population. However, these markers were only useful if there was a robust reference database of "ethnically" and regionally specified "African" DNA with which to match them. The few existing reference databases containing this information—including GenBank, an online database of publicly available DNA sequences that was first compiled in 1982—were not comprehensive enough to make reliable inferences about the regions in Africa from which the buried individuals might have come. For example, the GenBank database lacked DNA samples from some of the contemporary nations known to be part of active slave-trading regions, including Ghana, Angola, and Liberia.[28] Building on this incomplete foundation, Kittles began to compile a more comprehensive database.[29] As geneticists have long done, he also obtained samples from other scientists' private DNA databases, collaborating with molecular biologists with research sites in Francophone and Lusophone Africa, and collected his own samples from several African communities. Kittles's analysis of this subsample "indicated a strong West and/or Central African ancestral presence in the studied New York African Burial Ground individuals."[30]

In 1999, utilizing these techniques of analysis and database comparison on a third subsample that included forty-eight bone samples and two other hair and tissue samples, Kittles inferred that the "macroethnic affiliations" of forty-five of the burials he examined were in western and central Africa. The DNA sequences, suggesting the genetic diversity on the African continent, fell into the haplotype groups L1, L2, and L3—typically associated with African descent. The L2 haplotype that is indicative of West African Bantu speakers was present in close to 70 percent of the subset of burials that were genetically assessed. Kittles also deduced specific affiliations for some of the remains in Benin (the Fulbe people), Niger (Hausa), Nigeria (Fulani), and Senegal (Mandinka). (In 2000, after Kittles left the African Burial Ground research project, biological anthropologist Fatimah Jackson embarked on analysis of 219 of the remains at her lab at the University of Maryland and focused on correcting "the serious lapses in the existing database on African genetic diversity." Working with West African researchers, she began to compile the "first human databank in Africa.")[31]

These genetic insights about the burials were gleaned despite not only technical hurdles but also financial ones. Because the GSA funding did not cover DNA analysis, Kittles independently applied for a grant to support this research. Although he was not awarded the funding, his attempt to obtain it would further strain his relationships with the senior scientists on the project. For, according to Blakey, Kittles applied for these funds without informing the African Burial Ground Project's principal investigators of his plans.

Taken together, the researchers' findings yielded a dire account of the lives of enslaved persons in New York City, the second-largest slave port in the United States (after Charleston, South Carolina) and, in the colonial era, the place with "the highest proportion of slaves to Europeans of any northern settlement." In the eighteenth century, the "vast majority of Africans in New York were enslaved," having arrived either directly from the continent or via the Caribbean. Among this population, "premature mortality" was very high, including infant mortality rates almost twice that of the whites in the New York colony. More than 50 percent of the population died in childhood, and analysis of the remains showed evidence of severe anemia, malnutrition, and stunted growth. Slavery took a toll on the bodies of African men, women, and children that was apparent long after their death, in limb bones and bone joints that were "stressed . . . to the margins of human capacity."[32] As Blakey would describe in a 2003 interview, the slaves "were worked at the expense of fertility, at the expense of life; they were worked to death."[33]

The African Burial Ground research study formally concluded in the fall of 2003, with the ritual reinterment of the remains of the 419 African individuals whose graves had been excavated and the commemoration of their lives and those of the scores of others—estimated at between fifteen and twenty thousand in total—who had remained buried in the 6.6-acre cemetery.[34] But although the remains had been reburied, "public curiosity about this country's African-American past had been aroused by [this] New York experience," and these seemingly local concerns have come to have a significant afterlife in broader US political culture.[35]

AFRICAN ANCESTRY, INC.

Even before the results of the African Burial Ground study were published, hoping to build on the significant methodological breakthroughs he had engineered, Kittles embarked upon a plan to convert these methods of genetic analysis into a commercial enterprise, a proposal that produced an immediate schism among the researchers. Blakey, for example, publicly stated that it was "a questionable matter that a former researcher should take part of our program currently in development and do what he is doing."[36] Although today Fatimah Jackson serves on the board of advisors of Henry Louis Gates Jr.'s for-profit genetic genealogy company, African DNA, in 2003 she criticized Kittles's enterprise, declaring that it was "immoral to charge victims of slavery" for genetic analysis that might provide for them some indication of their ancestral origins.[37] Others later charged that the African reference databases that Kittles and other genetic-ancestry-testing entrepreneurs compiled were as of yet inadequate to make reliable inferences, being overrepresentative of some African populations and underrepresentative of others.[38]

Kittles's colleagues were also deeply bothered by his plans because they had imagined a different path for these technologies: Blakey and Jackson wanted the reference DNA database and genetic-ancestry-tracing techniques to be made available free of charge to members of the public curious about their ancestral links to Africa after the methods were more sound.[39] Community activists shared this vision. Ayo

Harrington, chairwoman of the Friends of the African Burial Ground, envisioned the site as including a museum of African history that would also contain a "DNA bank"—collected from the burial remains and stored at Howard University—that could be used by descendants "to determine their origins." Said Harrington, "If we could find one person who could one day go to that DNA bank, and it was determined that that person was a descendent, although we all are, it would just be something that folks would celebrate around the entire globe."[40] Even Blakey, Kittles's mentor-turned-critic, had articulated this potential for DNA analysis:

> The scientific research now underway constitutes yet another dimension of a long-standing human rights struggle among African Americans. That effort may relate directly to the conventions of the United Nations pertaining to human and group rights . . . [elucidating] slavery's impact on the lives of our ancestors and, by historic extension, its impact on living descendants. . . . We seek to restore knowledge of the African-American origins and identities that were deliberately obscured in the effort to dehumanize Africans as "slaves."[41]

Just a few years after the decoding of the human genome, even well-educated skeptics like Blakey thought it held the potential to transform the terrain of social justice and human rights.

Yet as archaeologist Warren Perry recalled in an interview with me, Blakey's aspirations for genetic ancestry inference were trumped by his concerns about its technical limitations and his deep disappointment that Kittles had "bought into black capitalism . . . [and] the mirage of commodification. So, Michael canned him, I mean quick! Michael told us, 'Listen we're getting rid of Rick Kittles. . . . He wants to go out and sell this.' . . . A lot of us were surprised because [Kittles] was doing decent stuff. But Michael said, 'That's not what this project is about; it's not the spirit of the project.' "[42]

Like Blakey, Jackson initially had a vision for the social potential of genetic analysis that did not extend to its commercial applications by scientists and lay people engaged in market collaboration. Reflecting on her work on the African Burial Ground several years later, Jackson surmised that the lesson learned was that "scientific effort (including genetic testing) must address the research issues of studied groups and not just the priorities of scientists."[43] She holds up the African Burial Ground study, in which collaboration between like-minded scientists and activists was mostly successful in steering the research, "as a prototype for future genomic initiatives, particularly among groups that have historically been victimized, rather than assisted, by genetic studies."[44]

Despite the scientific and professional hurdles he faced, in 2003, just four years out of graduate school, the maverick Kittles launched African Ancestry, Inc. (www.africanancestry.com) with Gina Paige, in the roles of scientific director and president, respectively. A Washington, DC, native, Paige attended Stanford University, where she completed a bachelor's degree in economics in 1988, following in the footsteps of her father, an economics professor. She would go on to earn an MBA at the University of Michigan. Paige then became a business strategist, conducted product development and strategy management for several major corporations, including the now defunct Sara Lee and Colgate Palmolive, while keeping a foothold in her own entrepreneurial pursuits.

Paige and Kittles met through Cynthia Winston, the former's cousin and a mutual acquaintance. As Paige recalls, Winston "knew that [Kittles] wanted to commercialize his research." While she is a member of Alpha Kappa Alpha, a black sorority, and joined the Black Student Union as a student at Stanford, Paige confesses that when considering the partnership with Kittles she did not necessarily share his African-centered political mission. "I was an integral part of the black community [on campus] socially, but not necessarily politically," she stated.[45]

But she knew immediately that Kittles had both a good idea and an important one. Both recognized that in order to succeed this start-up venture needed someone with Paige's experience. "I looked at it strictly from a business perspective. . . . For me, it was an opportunity to use my skill set to create and launch a product that had never existed before, for a group of people that I am passionate about. . . . That's what drove me to partner with Rick." For his part, Kittles had the technical vision but lacked business acumen. His efforts were "frustrated until I met Gina," Kittles acknowledged in an interview.[46]

When African Ancestry launched in 2003, there were only four other DTC genetic ancestry testing companies in the US market, Family Tree DNA (2000–); Gene Tree (2001–2013) and Relative Genetics (2001–2008), both owned by Sorenson Genomics; and Ancestry by DNA (2002–2009). One of the earliest direct-to-consumer genetic-ancestry-testing companies in the United States and the first targeted specifically at persons of African descent, African Ancestry by its own account has today tested more than 150,000 root-seekers in just over a decade of business.[47]

MatriClan and PatriClan are the brand names that African Ancestry gives to its mtDNA and Y-DNA test kits, respectively. (In 2013, the company added an autosomal testing kit under the name myDNAmix.) Kittles and Paige's African Ancestry is an information-age business—the exchange of a fee for service takes place online and through the mail. The company mails test kits to customers that contain the tools necessary to secure a DNA sample. The customer returns the sample to African Ancestry; it is then amplified and sequenced by the company's lab partner, Sorenson Genomics of Salt Lake City, Utah.[48]

The MatriClan and PatriClan tests draw on the distinctive properties of mitochondrial and Y chromosome to infer ancestral ties to current nation-states or cultural groups. Using these analyses, a consumer's DNA sample is matched against African Ancestry's reference database of genetic samples. Known by the brand name African Lineage Database, it is said to contain "over 30,000 indigenous African samples" from thirty countries and more than two hundred ethnic groups in Africa.[49] After several weeks, a customer will receive a results package that includes a printout of the customer's Y- or mtDNA markers, a "Certificate of Ancestry," and historical information about the African continent and the country with which the individual was affiliated.[50]

If the sample matches reference DNA at a set number of genetic markers (typically ten or more in the case of Y-chromosome short tandem repeat sequences, or hypervariable sequences in the case of mtDNA), this individual can be said to have shared a distant maternal or paternal ancestor with a person or persons in the database population. Making use of this form of analysis, a typical African Ancestry result informs a root-seeker that her mtDNA can be traced to a contemporary African ethnic group, such as the Mende people of present-day southern Sierra Leone. On the other hand, a male customer's results might trace him back to the Bamileke people of what is today known as Cameroon and, in some rarer cases, patrilineage may connect a black root-seeker to the continent of Europe rather than Africa.

African Ancestry's genetic-testing services would become the centerpiece of cultural practices through which repair and resolution of manifold injuries precipitated by racial slavery is sought.[51] From its use more than two decades ago in an excavation project that culminated in the "first National Monument dedicated to . . . Americans of African descent," genetic ancestry testing is now mobilized in efforts to investigate, adjudicate, and remember the history and extant consequences of the transatlantic slave trade.[52] As an "emergent" practice that may open up new ways of rendering or expressing enduring social concerns, genetic root-seeking goes beyond the politics of the past.[53] African Ancestry's slogan, "Trace Your DNA. Find Your Roots," has taken on many meanings as its services have been put to a range of purposes and functions beyond family history. The African Burial Ground project and African Ancestry together laid the groundwork for the reconciliation projects explored in the remainder of this book.

FOUR

The Pursuit of African Ancestry

All these years later, I find out it's Ghana. . . . What if it's true?

REVISITING *ROOTS*

Genetic genealogy testing aligns with an enduring human desire: the search for roots and identity.

The appeal of genetic ancestry testing cannot be understood without also understanding the backdrop of the specific example author Alex Haley provided about how this should be accomplished and what effects it might produce. Haley cowrote the autobiography of Malcolm X in 1965, the late activist's influential story of political transformation. Malcolm X's life account concludes with his pilgrimage to Mecca, the high holy city of Islam. Around the same time, Haley began work on a second book about race, *Roots: The Saga of an American Family*, which would reverberate with Malcolm X's narrative, and also entail a pilgrimage of sorts.[1]

Now known to be fiction based on fact, *Roots* was published in 1976—the year of the United States' two hundredth anniversary—to great fanfare and with tremendous critical and commercial success. Christened as "the most astounding cultural event of the American Bicentennial" by esteemed Civil War historian Willie Lee Rose—who would also take Haley to task in the *New York Review of Books* for sloppy historical research—the book's first two-hundred-thousand-copy print run sold out immediately upon publication. Millions of copies have been sold in the intervening decades the world over.[2]

Roots, for which Haley received a Pulitzer Prize, tells the story of Haley's colorful family genealogy, which he traces back to The Gambia. The story is framed as the author's "epic quest": his prodigious efforts across years and continents to uncover his family's past. In 1977, when Haley's work was transformed into a television miniseries, the story of his ancestors' trials, tribulations, and resilience held the country in rapt attention for eight days. The airing of *Roots* bested audience numbers for the inaugural television broadcast of *Gone With the Wind*, the previously most popularly watched show, and one that was notably also concerned with the formative role of slavery in US history.[3]

Haley's story came under scrutiny soon after it began to circulate, and was criticized for historical inaccuracies. He was accused of plagiarism on several occasions; one case would result in a settlement amount so large that it was effectively an admission of guilt. But these accusations did not present an obstacle to the story's power, and the narrative remains a commanding cultural symbol, national script, and racial allegory. British literary critic Helen Taylor, in an exhaustive essay on the book and television show, described their significance this way: "The impact of Haley and *Roots* has been profound. For African Americans, deprived for centuries of their ancestral homes and families, enslaved and exploited, denied basic human and civil rights . . . this book . . . offered a fresh perspective on their history, community and genealogy."[4] Writers who have followed the trajectory of Haley's *Roots* and written critically about it have hailed the work's cultural impact, despite its flaws. For Philip Nobile, for example, Haley was a "Colossus" with a "cultural halo" who "salvaged his lucrative career and preserved the myth of Kunta Kinte."[5] Taylor suggests that there has been a surprising public reticence surrounding accusations that Haley's work was marred. She ventures several reasons for these "surprising silences," including the fact that Haley was ensconced in elite networks. To borrow a contemporary phrasing, *Roots* was simply too big to fail. This was also the case because Haley's account of the Middle Passage and its consequences became an urtext—or primary narrative source—of African diasporic reconciliation for a generation of Americans. The story provided a narrative about slavery and its afterlives on the two hundredth anniversary of a nation that had never fully acknowledged its past. In place of a presidential apology for slavery, or a national discussion on racism, or the promise of reparations, we had *Roots*.

Haley started a social transformation in how we access and interpret the past. *Roots* generated excitement around family history; it encouraged the democratization of a practice that had previously been the provenance of the nobility. In the wake of the phenomenal success of Haley's book and miniseries, "root-tracing kits" containing family-tree templates and fill-in-the-blank genealogical charts on "imitation parchment" came on the market in the late 1970s; they were progenitors of today's genetic-ancestry-testing services.[6] Family-history research became a popular pastime for those seeking to discover unknown ancestors. *Roots* was the result of the author's efforts to uncover the mystery of his ancestral origins with clues garnered from Gambian griots, archival research, and his own genealogical imagination. Many have modeled their own ancestry pursuits on Haley's embellished account of his efforts to trace his familial lineage to Africa.

Genealogists of African descent frequently reference *Roots* when describing how their interest in family-history research was piqued. Many of the genealogists with whom I spoke—typically aged forty years or older, college educated, and predominantly female—were inspired by Haley's example as teenagers or young adults. The predominance of women in genealogical communities is consistent with the literature on "kinkeeping," the term coined by Carolyn J. Rosenthal to describe how the practice of maintaining family ties—through activities such as fostering communication between members or providing emotional and financial aid to them—was a form of gendered labor. Genealogists can be seen as fulfilling the role of kinkeeper in their families. With genealogical practices of prior times and of today, kinkeeping involves the work of connecting past and present kin with purposeful narrative.[7]

The experience of one of the genealogists with whom I spoke, Elisabeth (a pseudonym), is typical. I met Elisabeth, a computer scientist in her late forties, in an online community of black genealogists to which we both belong. Subsequently, I interviewed her at her home in the northwestern United States in 2004. She described the chain of events that had led her to become a genealogist and, some decades later, a genetic genealogist, and waxed nostalgic about a presentation by Haley at her midwestern high school that had stimulated her interest in genealogy:

> Haley came to my high school in 1970. This was before *Roots* came out. He had a *Reader's Digest* article about it out and he was on the road just telling everyone about how he traced Kunta Kinte. And I was in ninth grade and I just sat there mesmerized. . . . Actually, I have a copy of the tape [of Haley's presentation]. I got in contact with my old high school civics teacher, out of the blue, last year. And he says, "You know, I was going through stuff and I found this old Alex Haley tape. I didn't know what to do with it—would you like it?" Of course! And it's phenomenal! . . . It was just a fascinating talk; it really was. That's when I got bit by the genealogy bug.

Elisabeth's friend Marla (a pseudonym) expressed similar sentiment about Haley's influence when I met with her. Although she made a start at genealogical research in the 1960s, following the death of the eldest member of her extended family—its kinkeeper—it was not until a decade later, when she attended a lecture by the author at a local community college, that she became serious about the endeavor. This encounter impressed upon her that a nonspecialist researcher could employ insurance records, land deeds, slave-ship manifests, and family history libraries to trace her roots to Africa. As she explained to me, "It was interesting to hear [Haley] talk about . . . going to the Mormon temple and going to Lloyd's of London and all of that. I never figured that I would have access to those kinds of records . . . I never ever thought that the average person could have accessed it. So I never anticipated being able to . . . go back to slavery."

Until recently, for persons of African descent and others, pursuing one's family history has typically entailed genealogical excavation of the type depicted by Haley. Although root-seeking methods have evolved, Haley's influence remains; the example of his project established an expectation among a generation of readers and viewers in the United States and abroad that recovering ancestral roots was not only desirable, but possible.[8]

Why this avid interest in genealogy today? The fact of Haley's *Roots*—the powerful testimonies that I heard from Elisabeth, Marla, and others notwithstanding—by itself is an incomplete explanation. Haley's narrative did provide an inspirational account of African American genealogy, but it also prompted an international conversation on slavery's bequest to us.

And notably, it supplied a narrative for black life. Noting the black-power-era context of Haley's book, historian David Gerber commented on the striking similarity between Haley's ancestor-protagonist in *Roots*, Kunta Kinte, and his coauthor Malcolm X, noting that both fulfilled collective emotional need for inspirational models "of strength, dignity, and self-creation in a hostile or, at best, indifferent White world."[9] Gerber's insight points us to the fact that *Roots* says as much about the time at which it was written as it does about the past. *Roots* is, as Taylor argues, a book that resounds at a personal familial level yet also invites historical reckoning, social identification, and political resonance.

Similarly, the interest in genetic ancestry testing needs to be understood as reflecting the particular concerns of this moment. As the French sociologist Maurice Halbwachs argued in his important work on collective memory, we conjure the accounts of the past we need in order to tackle issues of the present.[10] With this in mind, we can understand how practices of genetic genealogy and family history that are anchored in the past become a form of contemporary racial discourse. The popularity of DNA testing is a symptom of *Roots*' unfulfilled promise, and it should therefore come as no surprise that it is sought to balance the ledger of a racial economy of inequality.

THE RIGHT TOOL FOR THE JOB: ROOT-SEEKING STRATEGIES

Despite Haley's heroic, if flawed, example, few African Americans are able to fill in the contours of their past as he did, owing to the decimation of families that was a hallmark of the era of racial slavery and the dearth of records from this period. As a consequence, genetic genealogy testing, which is now broadly available and also less taxing—and, owing to the social power of DNA, seemingly more authoritative—than conventional Haley-esque genealogical research, holds considerable appeal for many root-seekers.

Genetic genealogy testing emerged from techniques developed in molecular biology, human population genetics, and biological anthropology.[11] Direct-to-consumer genetic testing was first available in the United States in 2000 from Family Tree DNA, a pioneer in this field that remains an industry leader.[12] When African Ancestry was launched just a few years later, the company was as notable for joining the cutting edge of a new social and technical practice as it was for its niche mission and customer base. By 2004, five other American companies had joined the ranks of African Ancestry and Family Tree DNA. By 2010, thirty-eight companies worldwide offered an array of genetic-ancestry-testing services, with twenty-eight of these based in the United States.[13]

The companies that sell DNA analysis for genealogical purposes offer three principal forms of analysis for which they create brand names, such as African Ancestry's MatriClan and PatriClan. Rather than provide companies' brand descriptions and for the sake of analytic clarity, here the tests will be classified by what information they provide as an end result to the consumer, because the forms of social orientation that the test results suggest are of primary importance to root-seekers. The genealogists I have spoken with purchased particular genetic tests in order to fulfill distinctive genealogical aspirations, such as corroboration of a multicultural background or association with an ethnic community.[14] The three broad classes of DTC genetic ancestry services can be categorized as spatiotemporal analysis, racial-composite analysis, and ethnic-lineage analysis.

With spatiotemporal testing, a consumer's DNA sample is classified into a haplogroup (sets of single nucleotide polymorphisms [SNPs] or gene-sequence variants that are inherited together) from which ancestral and geographical origins at some point in the distant past can be inferred. The result orients a consumer in space and time but does not provide identity per se. This form of analysis was made possible by the ambitious Y-DNA and mtDNA mapping research that resulted in theories about the times and places at which various human populations arose. Y-DNA and mtDNA have distinctive sequence combinations; similar sequences can be classified into broad groups, called *haplogroups*. Human population geneticists have devised a system of letters and numbers to identify the region of one's ancestors, and also the time in history (hundreds of thousands of years ago) during which they would have migrated from Africa. Family Tree DNA, a forerunner in American genetic ancestry testing, supplies customers with haplogroup information, as does National Geographic's Genographic Project. An inferred match with the mtDNA-derived L2a haplogroup—a designation shared by some of Venture Smith's descendants—suggests that one's ancestors lived in Africa approximately sixty thousand to eighty thousand years ago.[15] In short, spatiotemporal analysis offers "deep" ancestry results that open a window onto the geographic past of ancestors who may have dwelled in a time and place far removed from where root-seekers presently abide.

Among the more spectacular claims of genetic ancestry testing is the ability to infer not merely where we come from but *what we are*, in the most essentialist sense. These tests, which I classify as racial-composite analysis, claim to ascertain the percentage of three of four supposedly "pure" racial groups. In contrast to spatiotemporal and ethnic-lineage analyses, which rely on mtDNA and Y-DNA, this genomics testing involves the analysis of nuclear or autosomal DNA, which is unique to each person (identical twins excepted, although this is now being debated) and consists of the full complement of genetic information inherited from parents. A DNA sample is compared with panels of proprietary SNPs that are deemed to be "informative" of ancestry. Algorithms and computational mathematics are used to analyze the samples and infer the individual's "admixture" of three of four statistically constituted racial categories—African, Native American, East Asian, and European—according to the presence and frequency of specific genetic markers said to be predominate among but, importantly, not distinctive to, each of the "original" or "pure" populations.[16]

This form of analysis was first developed by the DNA division of DNAPrint Genomics. When this Florida-based company launched in 2002, it offered the "first *genomic* ancestry test"; that is, a test based on complete autosomal DNA.[17] A subject of this racial composite testing might learn he is estimated to be 80 percent African, 12 percent European, and 8 percent Native American. DNAPrint Genomics went out of business in 2009, but other companies offer similar services, including African Ancestry's myDNAmix and 23andMe's Ancestry Painting. This type of analysis proves useful to those who think that what we understand as racial groups are self-contained and, therefore, that mixture can be ascertained. However, it offers little guidance about one's geographic or ethnic origins other than in the broadest sense. As

African Ancestry cautions its customers on its website, "YOU WILL NOT LEARN COUNTRIES OR ETHNIC GROUPS."

With ethnic-lineage testing, an individual's DNA is searched against a genetic-ancestry-testing company's reference database, which is in most cases proprietary, thus the claims made using it cannot be independently verified. (This is also true of the other types of analysis; DTC genetic testing companies hold data and algorithms as trade secrets.) A match between the sample and the reference DNA or shared haplotypes suggest a shared, distant maternal or paternal ancestor. Most companies offer this type of testing. A typical ethnic-lineage result may inform a test-taker that her mtDNA traced to the Mende people of contemporary southern Sierra Leone. Or a male customer could be inferred to be ancestrally linked to a group of male genealogists who share his surname and Y-DNA profile. African Ancestry's analyses might thus be regarded as ethnic-lineage instruments through which an undifferentiated racial identity is translated into African ethnicity and kinship. By linking blacks to inferred ethnic communities and nation-states of Africa, African Ancestry's service offers root-seekers the possibility of constituting new forms of identification and affiliation.

Each of these three types of tests thus offers a different window onto the past and, as Halbwachs would remind us, also a distinct vantage on the present. Root-seekers demonstrate their preferences for genetic information in the form of the testing they select and purchase. The usefulness of test results depends on the perspective of the root-seeker and the particular questions he or she seeks to answer through genetic genealogy analysis. In my encounters with genetic genealogists, one of the more important insights I gained is that root-seekers' preferences are shaped by the problems to which they are applied. It should also be noted, however, that many of the test-takers I met used more than one type of genetic genealogy analysis, typically to compare results received from different companies or obtain new information from a company from which services were purchased previously (for example, when a company releases a more robust form of test that employs more markers or has added a significantly larger number of samples to its reference database).

GENEALOGICAL ASPIRATIONS

Identity and self-making are primary ambitions for genetic genealogists. Questions and desires, and not "pure" science alone, set the terms for how a personal reconciliation project—the pursuit of African ancestry—is carried out. Consumers come to DNA testing with genealogical aspirations: with particular questions to be answered; with mysteries to solve; with autobiographical narratives they want to complete. These aspirations may precondition how genetic test results are received by consumers, and may prompt an uneasy negotiation with the information supplied by genetic genealogy companies.

Like those family members and researchers interested in the genetic ancestry of Venture Smith, the root-seekers I encountered over the years have come to invest more confidence in the ability of DNA analysis to augment their family history research in new and exciting ways. These root-seekers tend to know far less about their enslaved ancestors than Venture Smith's kin, but they share the same genealogical aspirations for ethnicity voiced by New York African Burial Ground activists and researchers (sometimes alongside other ancestral goals). They are united in the aspiration to African ancestry as an end in itself. They are also engaged in a form of reconciliation, for genetic ancestry testing is, in an elemental sense, always as much about the reconstruction and reunion of the family and community as it is about the individual.

While today's popular genealogy television programs would lead us to believe that root-seekers take up wholesale the information provided to them by genetic ancestry tests and accept it unconditionally, something far more complex is at play. Genetic genealogy tests are deemed reliable to the extent that they are useful for consumers' myriad aims; for many, this involves strategically marshaling the data. Some use their genetic results as usable narratives that open up new avenues of social interaction and engagement, for example. It is through these sorts of negotiations that contemporary racial politics have begun to move into the *terra nova*—if not the *terra firma*—of genetic genealogy.

Racial composite testing has proved unsatisfactory to some root-seekers who want to re-create Alex Haley's *Roots* journey in their own lives. Although composite testing analyzes an individual's full genome, its results nevertheless lack specificity and usefulness for some users. This was the case for one genealogist I spoke with, an African American woman I will call Cecily. Attractive and about fifty years of age, she wore her hair in short twists and sported a relaxed linen outfit on the day we met at an AAHGS meeting, striking up a conversation after one of Kittles's genetic genealogy presentations. As we sat near the display booth of the African Ancestry company, from which Cecily had previously purchased ethnic-lineage analysis, I asked whether she planned to pursue racial-composite testing as well. In response, she declared, "I don't need to take that test. We're all mixed up. We know that already."

Somewhat similarly, spatiotemporal testing results may be deemed too remote by some root-seekers, as was the case with Marla. A black Californian, Marla is in her late fifties and retired from a job with the US Department of Defense. With her salt-and-pepper Afro, impressive knowledge of many subjects, and precise language, she put me in mind of interviews with the late novelist Octavia Butler I'd watched over the years. In addition to the genealogy chapter she leads with Elisabeth, Marla also moderates an Internet forum dedicated to discussion of DNA testing for genealogical purposes and has purchased several tests. An mtDNA test purchased from African Ancestry matched her with the Tikar people of Cameroon. As I have found is frequently the case, Marla's initial testing experience stimulated further curiosity about her ancestry rather than satisfying it fully. She then purchased a racial-composite test for herself and also paid for three family members to have ethnic-lineage testing through Trace Genetics (a company known for its large database of Native American reference samples, which was purchased by DNAPrint Genomics in 2006 before the latter company ceased operations in 2009; it is now defunct).

For another round of testing, Marla sought to find out more about the maternal line of her deceased father. As a seasoned genealogist, she knew that this information could be accessed if she had a paternal second cousin's DNA analyzed. In an e-mail exchange between Marla and me that followed from a conversation we had at her home, she detailed Family Tree DNA's spatiotemporal analysis of her cousin's genetic sample:

> The mtDNA of my 1st cousin's daughter (paternal grandmother's line) traced to "Ethiopia" and ±50,000 years ago. It is Haplogroup L3 which [according to the information provided by the company] "is widespread throughout Africa and may be more than 50,000 years old." Her [the cousin's] particular sequence "is widespread throughout Africa" and has its "highest frequency in West Africa."

Marla stated that the results were "deeper" than she had wanted and referred to ancestry "far before the time that I am interested in." She expressed her frustration that these genetic genealogy test results did not provide her with more information than she might have surmised on her own:

> Huh???? Ethiopia? West Africa? Didn't just about everybody outside Africa come through the Ethiopia area 50,000 years ago? Maybe I'm off by a few thousand years. . . . These kinds of results are meaningful for those tracking the worldwide movement of people (like the National Geographic study), but not really meaningful to me in my much narrower focus.

Marla concluded our exchange by informing me of her plan to send these results to the African Ancestry company for reinterpretation and comparison against its African Lineage Database.

As my interactions with Cecily and Marla revealed, successful genetic-ancestry-test outcomes are those that offer root-seekers what they deem to be a serviceable account of the past. For Cecily, racial-composite analysis would merely confirm the "hybridity" she knew existed given the history of racial slavery in her family. To her mind, this form of genetic genealogy testing provided information that was neither novel nor useful. Given Marla's stated aim to try to derive ethnic-lineage results from her spatiotemporal ones, the "much narrower focus" that would be "really meaningful" to her would apparently take the form of a genetic genealogy result that affiliated her with an African ethnic group and possibly a present-day nation, thus fulfilling the genealogical aspiration that was seeded when she attended a presentation by Haley three decades prior. Taken together, Cecily's indifference toward racial composite testing and Marla's preference for ethnic-lineage analysis suggest that not just any scientific evidence of ancestry will do. Genetic root-seekers strategically seek out the right tools to fulfill the genealogical work at hand.

Genetic genealogy test results may challenge not only the genealogist's prior expectations but also other evidentiary bases of self-perception and social coherence. As Marla's response to her spatiotemporal result implies, and as I elaborate below, root-seekers are not only judicious about the types of genetic genealogy tests they purchase, but deliberative in ascertaining the significance of their results.

DNA SPILLOVER

I attended a symposium on race and genetics at a large public urban university in the Midwest in the fall of 2003. It was a small, interdisciplinary gathering of scholars and included presentations by social scientists, geneticists, and bioethicists, among others. The audience consisted mostly of symposium presenters, but also included interested faculty affiliated with the university and members of the public, who sat in on discussions for short periods of time throughout the day. A number of non-academics were on hand for an afternoon presentation by Kittles, who was, at the time, a researcher at the National Human Genome Center at Howard University, in addition to serving as the scientific director of his recently launched genetic-ancestry-testing company. In his talk, Kittles discussed the scientific research and sociocultural assumptions behind the ethnic-lineage analysis his company had begun offering several months prior. During the presentation, I sat next to a middle-aged African American woman whose steel-toed work boots and navy cotton jacket emblazoned with Teamsters Union patches placed her in a somewhat different category than the academics in attendance, who, like me, were dressed in business-casual attire and hunched over our notebooks and laptops. While Kittles spoke, the woman nodded enthusiastically in assent and, from time to time, looked over to me seeking mutual appreciation of the geneticist's presentation. I smiled and nodded in return. This silent call-and-response went on for several minutes, when at one point she leaned in and whispered to me that she had "taken his test."

At the conclusion of Kittles's presentation, the woman (Pat) and I continued our discussion of her experience with African Ancestry's genetic genealogy service as she walked with me through the labyrinthine campus. She spoke of her interest in conventional genealogy and of recent events that had prompted her to use DNA analysis to trace her African roots. Pat (a pseudonym) shared that she was a long-standing member of the AAHGS and of two other genealogical societies. For almost thirty years, this root-seeker had assembled archival materials, reminiscences, oral history, and linguistic clues from family members. This evidence led her to deduce that her family's maternal line may have descended, in her words, from "the Hottentots" (or the Khoisan of southern Africa). Despite some success with her genealogical research by traditional means, Pat had not been able to locate a slave-ship manifest or definitive documentation of her African ancestry. She told me that, as a result, "some missing links" remained to be uncovered.

I asked if she thought genetic ancestry testing was reliable. Pat replied, "I've seen people let off jail sentences based on DNA. . . . I'm not question[ing] about DNA . . . given my experiences [working in the lab], there is no reason to doubt the technology." Prior to Pat's employment at the university, she processed forensic evidence for a police department crime lab in the same city. This work experience bolstered her confidence in African Ancestry's product.

DNA spillover occurs when an individual's experience with one domain of genetic analysis informs his or her understanding of other forms of it or authorizes its use in another domain; this was the case with Pat, who drew an association between criminal forensic genetics and genetic genealogy. A similar dynamic was at play for a genealogist I will call Ruth, who told me that she gained a greater understanding of the inheritance of disease following her genetic-ancestry-testing experience. "We think breast cancer runs in our family," she explained. "Now that I understand my African Ancestry test—the difference between the mother's line and the father's line and all that—I have a better sense of what the genetic counselor at my doctor's office was telling me." DNA spillover has its upsides.

But, as Pat's comments partly suggest, it may also cause us to be less critical of the Venn diagram nature of the social life of DNA than is warranted. With black men and women comprising close to half of the two million people incarcerated in the United States, the genome era coincided with the cresting of racialized mass incarceration. One of the predominant ways that many African Americans encounter DNA analysis is through the criminal justice system, for the purposes of both exoneration *and* conviction.

In 1992 Peter Neufeld and Barry Scheck, faculty members at the Cardozo School of Law in New York City, initiated the Innocence Project. Neufeld and Scheck proposed to use DNA analysis to exonerate wrongly convicted persons and, simultaneously, shine a light on biases in the criminal justice system. The Innocence Project soon spread to other institutions and cities. Within a decade of its founding, there were forty-one similar projects nationwide. To date, 330 men and women have been released based on this advocacy. The bittersweet success in revealing injustice led scores of states to change their laws in order to make post-conviction forensic genetic testing more readily available. But the laudable work of the Innocence Project poses a continuing threat to the legal status quo: "[P]ostconviction review is dangerous to incumbent officials because of the possibility that it will reveal errors by individuals and the system," criminologist David Lazer explains.[18]

The reticence of some authorities with respect to this legal advocacy, however, has been overshadowed by the dramatic press accounts, television programs, and films recounting the powerful stories of incarcerated innocents who have received some small measure of justice through the use of DNA. It is DNA's liberatory potential that Pat had in mind when she spoke to me of her faith in genetic technologies. The social power of DNA (to exonerate) increased her confidence in genetic genealogy's powers of ancestral identification. Yet an unanticipated outcome of the urgent work of the Innocence Project is the "halo effect" on DNA analysis both for good and for naught, including the growing practice of collecting DNA from individuals upon arrest for even minor offenses (following the *Maryland v. King* Supreme Court ruling permitting this practice), and the assumed unassailability of "DNA fingerprinting" in criminal cases.

Pat's resolute belief in genetic analysis paralleled her faith in African Ancestry's chief scientist. When I asked if she had had any apprehensions prior to purchasing her MatriClan test, she unequivocally replied,

"I trust Dr. Kittles." This comment is of signal interest in light of the fact that, owing to a legacy of racially segregated healthcare and experimental exploitation, including the notorious Tuskegee syphilis study, African Americans can be distrustful of medicine and scientific research. This distrust has been shown to negatively impact blacks' health-seeking behaviors and to create a disincentive for African Americans to participate in clinical trials. In contrast, the growing popularity of African Ancestry's services demonstrates that this understandable skepticism may be assuaged by the presence of Kittles, who wields authority as a scientific researcher and over the last decade has become a widely known figure.

Kittles is among the most well-known molecular biologists, and perhaps the best-known African American geneticist, in the United States—a reputation he has burnished through extensive media coverage, scholarly publications, and institutional associations. Kittles appeared, for example, in the 2003 BBC documentary *Motherland: A Genetic Journey*, as well as in Henry Louis Gates Jr.'s PBS documentary *African American Lives* in 2006, in which African Ancestry's services were used to trace the genealogy of celebrities including Oprah Winfrey and Morgan Freeman, among other notables. He has also been featured on *Good Morning, America*; *The Morning Show*; and *60 Minutes*. Scores of newspaper and magazine articles, including in the *New York Times*, *Time*, the *New York Daily News*, *Black Enterprise*, *Wired*, *Fortune*, and the *Los Angeles Times*, have included commentary from Kittles, while his numerous articles in the area of human variation and genetics have appeared in such leading journals as the *American Journal of Human Genetics*, *Science*, the *Annals of Epidemiology*, and the *American Journal of Public Health*. From 1998 to 2004 Kittles was an assistant professor of microbiology at Howard and a director of the molecular genetics unit at that institution's National Human Genome Center, and he has also held positions at Ohio State University, the Cancer Research Center at the University of Chicago, the University of Illinois at Chicago, and, at present, the University of Arizona College of Medicine. Even as Kittles was solidifying his position as an expert on DNA and ancestry, his "hard" scientific research on the genetic determinants of prostate cancer—a disease that disproportionately afflicts African American men—helped legitimize his forays into what is regarded by some as the "softer" science of genetic genealogy. At the same time, his renown as a scientist and his involvement with cutting-edge medical genetics research lend authority to his commercial genetics enterprise.

Kittles's combination of scientific prestige and cultural competency is an unmistakably important aspect of the appeal of African Ancestry's genetic genealogy testing and of consumers' faith in the results it supplies to its customers. In his guise as a genealogist colleague who shares his customers' desires for ancestral reckoning and their reservations about the potential for misuse of DNA analysis, Kittles establishes genetic genealogy testing as a legitimate and safe practice for African American root-seekers. Many root-seekers are as compelled by Kittles as they are convinced by genetic science. Kittles is a fellow traveller of African descent, familiar with the dialects of black experience.

GENEALOGICAL DISORIENTATION

Pat told me that she proceeded to purchase an mtDNA test from African Ancestry following a persuasive pitch by Kittles at one of the three genealogy club meetings she regularly attends. Asked to recall her feelings as she awaited her test results, she responded, "I didn't know what to expect . . . it's like rolling a lottery thing; okay, this is where it landed." Prior to testing, Pat had some information about her ancestry, but no preconceptions about her ethnic "match." A comparison of Pat's DNA with the company's African Lineage Database did not place her maternal line in southern Africa as she anticipated from family lore and earlier research. Instead, she was associated with the Akan, a large ethnic group of Ghana and southeastern Cote D'Ivoire that includes the Asante, the Fante, and the Twi, among others. Pat's results included "my genotype" printed out on paper, "a letter of authenticity from the lab," and "a certificate saying I was Akan." However, these authoritative artifacts did not leave Pat feeling fully settled about her genetic ancestry. She recollected, "I felt numb, blank. [I've] been doing genealogy since 1977. I grew up with knowledge of Hottentot . . . all these years later, I find out it's Ghana." She added, after a pause, "What if it's true?"

Pat's uncertainty about her results increased several weeks later, when she learned that other members of her genealogy club reported receiving the same ethnic match from African Ancestry as she did—Akan. These results were likely accurate in the statistical universe of proprietary gene-sequence variants used by the company and in light of substantial historical research showing that current-day Ghana and other western African countries were key nodes in the transatlantic slave trade. Nevertheless, the preponderance of similar ethnic lineage findings among her genealogist colleagues, and the inconsistency of her genetic result with the family genealogy she had assembled by conventional means, led Pat to conclude that "we still technically don't know who we are."

It is worth dwelling for a moment on the word "technically," because the scientific credibility of genetic genealogical analysis can be fraught. The mtDNA and Y-chromosome DNA analysis that is widely used to infer ancestry is beneficial in that it passes mostly unchanged among mothers and children and fathers and sons, respectively; however, it provides less useful information about the breadth of one's ancestry. Each genetic lineage is estimated to provide less than 1 percent of one's total ancestry. Put another way, these analyses follow ancestry of a single individual back ten generations and more than one thousand ancestors, yet matrilineage and patrilineage testing only offer information about a portion of these. If we think of one's ancestry as an upside-down triangle, these forms of ancestry tracing follow the lines to the left and right of the triangle point, but offer no details about the shape's filling.

Purveyors of genetic genealogy testing claim that their services trace or reveal otherwise unavailable information about ancestry and ethnicity. However, at present, matching a consumer's DNA against genetic databases comprising samples from contemporary populations, as genetic-ancestry-tracing companies do, cannot establish kinship with certainty; ethnic-lineage analysis does not associate a root-seeker with specific persons at precise locations in time and space. Also, owing to both technical limitations (e.g., mtDNA and Y-DNA tests compare a consumer's genetic sample to a selective proprietary sample database and analyze a small percentage of a test-taker's DNA; and provide probabilistic outcomes) and historical dynamics (e.g., racial and ethnic identities are sociocultural phenomena and the unpredictability of human migration patterns suggest that contemporary social groups cannot be easily correlated with earlier ones), the associations inferred through genetic genealogy are necessarily provisional.[19]

What's more, contemporary populations cannot be technically delimited in any absolute sense. University of Massachusetts geneticist Bruce Jackson, who endeavored to help Bettye Kearse use DNA analysis to determine whether she is related to President James Madison, is, perhaps surprisingly, a strident critic of genetic ancestry testing. As he explained to a reporter, "One of the greatest myths . . . is that someone can link you to an African tribe. . . . There are thousands of ethnic groups in Africa, and there are only a handful that these people are studying. So how can you take a person and link him to a tribe?"[20] He noted in the same interview that there are more than 250 ethnic groups in the nation of Nigeria alone, but few of these have been studied or incorporated into DNA biobanks or databases. Jackson's observation prompts reconsideration of the claims African Ancestry makes for its proprietary database, which contains samples from more than 200 ethnic groups; according to Jackson, this number does not even exhaust the potential ethnicities of a single West African country. To their credit, African Ancestry and other genetic genealogy purveyors have focused their efforts on obtaining DNA samples from the areas most heavily trafficked during the transatlantic slave trade, since it is this information that is of primary importance to black root-seekers. Mindful of the company's critics, Paige underscores African Ancestry's broader mission: "It's not a perfect science, but our goal is to give people some sense of place prior to the period of slavery."[21] Given the infeasibility of collecting DNA samples from across the continent of Africa, let alone the globe, Michael Darden, an African Ancestry spokesperson, put a finer point on it in the *Washington Post:* "Knowing that is better than nothing."[22]

Additional limitations to the information supplied by genetic-ancestry-testing services involve the subjective nature of the enterprise. How ancestry is defined and established is in some regards unique to each entity. Companies use proprietary databases and algorithms that are not made available to the public, other scientists, or business competitors. At the same time, a gene-marker sequence of mtDNA that is used to determine maternal lineage may "not exactly match one of the previously published haplogroups" or, as was the case with Pat, the designation may be found in numerous geographic locations and among several contemporary populations.[23]

Pat's use of the word "technically" in her estimation that "we still . . . don't know who we are" indicates that although she had expected genetic genealogy to supply her with roots in Africa, the results did not fully convince her. However, given her prior positive assessment of genetic testing, her reference to technical uncertainty may indicate her discomfort with conceptualizing family history as a technical matter. Her words intimated that the genetic-genealogy-testing experience produced a lack of orientation, and more particularly, "genealogical disorientation" as an affect ("I felt numb"), and as an effect of her misgivings about its reliability ("What if it's true?"). Pat also feels "blank." In her search for family, she has lost the familiar. For Pat and others with whom I have spoken, the receipt of genetic facts about ancestry opened up new questions about identity and belonging rather than settling them absolutely.

Since receiving her test results, Pat endeavored to reorient herself using the new information she received. She has begun a friendship with a Ghanaian neighbor and has embarked on research into the history and culture of the Akan. More recently, Pat has begun to explore the possibility of having roots in West Africa. Yet her genetic ancestry test has taken on deeper significance, not because of confidence in mtDNA analysis necessarily, but because of her own efforts to resolve her genealogical disorientation.

The growing appeal of Pat's Akan-ness was powerfully illustrated in the following account she shared with me: at a community Kwanzaa fair, Pat was faced with a purchasing decision that revealed her vacillating ethnic identity. Coming upon an African immigrant flag-vendor as she strolled through the fair, Pat was confronted with two symbols of her ancestral roots and putative nationality. She inquired about the significance of a flag with three color fields of red, black, and green. The vendor replied that it was a "general flag," indicating that it was a Pan-African flag, which symbolized the African diaspora rather than a specific nationality or ethnicity. Pat responded, "My DNA said I came back as Ghanaian. I don't need the red, black, and green." The woman replied, "Now you know, so you don't need just a plain flag anymore." She concluded that "if anything has changed [about how I perceive myself], it's that I bought my first Ghanaian flag last year." In this exchange, Pat's testing experience emboldened her to invoke her biology ("my DNA said I came back as Ghanaian") when offered an undifferentiated symbol of Africa by the vendor. She then asserts that she may not "need" the Pan-African flag. However, it is the African vendor's not uninterested response, "Now you know," that endorses and authenticates Pat's claim to Ghana, leading to the purchase of a symbol of her possible "home." To be sure, like other genetic root-seekers, Pat exercised choice in the interpretation of her test results. She had the freedom to link her ethnic-lineage testing to her purchase of the Ghanaian flag. Yet her exchange with the vendor suggests that her choice is somewhat constrained by the African flag vendor's cultural authority, just as it is by African Ancestry's authenticating but selective database. Although this social interaction authorized Pat's genetic affiliation with the Akan, her opinion of her family origins nevertheless remains in flux. Now when asked by others about the outcome of her root-seeking pursuits, Pat admits to answering "Akan" and "Hottentot" interchangeably.

AFRICAN IDENTITY AND THE MOTHERLAND

As described in the subtitle of Haley's book, it concerns the "saga" of "an American family." This account was a profoundly African *American* one. But Haley's deep delving into his family's past also resonated with a worldwide audience because the dynamics he described touched nearly every corner of the earth. Although racial slavery had especially severe effects in Liverpool and Kingston, Cape Coast and Charleston, Salvador de Bahia and Gorée Island, its shadow veils the globe. The dispersals of the transatlantic slave trade have created a diaspora of African root-seekers.

In 2004 I attended a symposium on race and genetics at the London School of Economics. One of the conference participants was Neil Cameron, executive producer of the 2003 BBC documentary *Motherland: A Genetic Journey*, who invited me to a gathering of the "Motherland Group," black Britons who participated in the study on which the documentary was based. (Many of the genetic genealogy reality-television shows that would begin to appear in the United States in a few years' time drew on *Motherland* as a model.) Study participants had volunteered DNA samples in exchange for the opportunity to have their ancestral links to Africa scientifically inferred. The Motherland Group first convened in March 2003, shortly after the premiere of the television show. According to Arthur Torrington, a participant in the study and leader of the group, members came together to discuss the "pros and cons of the testing," because, as he expressed it, the test results were but "the beginning of a journey; there is much more to this thing." Cameron and his production partner, Archie Baron, subsequently arranged presentations to the group by genetics experts. As Cameron said to me, group members "wanted to be able to talk to each other about the experience and learn more about the science behind the study."

This meeting served as a forum for the coproduction of biological and social identities, for the making of what anthropologist Paul Rabinow has called "biosociality"—shared medical conditions and genetic predisposition to disease or disability.[24] However, participants did not gather on this basis, but rather assembled to explore what biosociality might result from their testing experience. This was *bios* put to the task of creating a very particular kind of sociality: the possibility of affiliation based on ethnic-lineage and racial-composite DNA analysis.

Although DNA samples from 229 persons were analyzed for the Motherland study, the documentary featured just 3 participants, chosen by the producers for their telegenic appeal and for the dramatic potential of their narratives. Cameras accompanied Jacqueline Harriott as she traveled to Jamaica to explore her Caribbean heritage and await the results of her racial-composite analysis.[25] *Motherland* viewers traveled as well with Mark Anderson and Beaula McCalla as they were transported to their supposed, respective pre–slave trade "motherlands" of Niger and Bioko Island (an island of Equatorial Guinea, a nation composed of mainland territory and five islands) for a dramatic "reunion" with their lost kin.

Mark Jobling—the University of Leicester geneticist who was

involved in research in the 1990s to determine the paternity of the children of Sally Hemings, the woman owned by Thomas Jefferson—analyzed Anderson's Y-DNA. The results gave Anderson quite a surprise. As shown in the documentary, he is stunned to learn that his Y-chromosome traces to Europe rather than Africa. His genealogical disorientation is raw and apparent. Because *Motherland* was a pioneer of genetic genealogy reality television, before the narrative arc and ritualized performance of these programs had been codified into a script as it is seen today in shows like *Who Do You Think You Are?* (also borrowed from British television), there is a compelling authenticity to Anderson's reaction.

Like other of the African American genealogists I interviewed, such as Marla, Anderson wanted results that fit with his present self-understanding. Also like Marla, he turned to Rick Kittles for additional assistance. As the documentary depicts, Anderson sees Kittles as a trusted voice and perhaps even a hero. Kittles is shown explaining to him that more than 20 percent of Y chromosomes of men of African descent will trace to Europe. The black geneticist then provides Anderson with the information—the usable past—he was clearly seeking, a link via mtDNA analysis to the people of contemporary Niger. Mark is elated by this news. His embrace of results that better aligned with his self-conception reveals the extent to which these genetic ancestry techniques are valuable to root-seekers in fulfilling genealogical aspirations that can be deployed in the world. The utility of this scientific information is depicted in *Motherland* as we see Anderson taken "back to Africa" to meet with some of the Kanuri community, his inferred ethnic group in Niger.

Back in London, Beaula was in attendance at this meeting of the Motherland Group, which featured a specialist in African migration, human evolutionary geneticist Martin Richards from the University of Leeds, as the day's speaker. His presentation outlined the theoretical and technical assumptions on which the study participants' genetic genealogy results rested, and cautioned attendees about the limitations of mtDNA and Y-chromosome analyses for the purposes of determining ancestry. Specifically, Richards warned attendees that genetic genealogy does not link a consumer to ancestors at a specific location and historical moment. He stressed as well that the DNA databases on which genetic genealogical tests rely are incomplete, because they contain too few samples from too few sites in Africa to make robust claims.

On these points, Richards was repeating arguments first aired in an editorial critique of the Motherland study that he published in the *Guardian* newspaper one year earlier. Entitled "Beware the Gene Genies," his opinion piece revealed that the genetic marker used by Cambridge University geneticist Peter Forester to link Beaula with the current-day Bubi people of Bioko Island, Equatorial Guinea—called a "rare marker" in the documentary—was also found thousands of miles away. Richards wrote that "a glance at the published mitochondrial database shows that Beaula's variant is also found in Mozambique."[26] He continued, noting that "a huge area of central and southern Africa that provided more than a third of all victims of the slave trade is still unsampled. Beaula's maternal lineage could have come from anywhere in that region."[27] I was later told by producer Cameron that Motherland Group members were familiar with Richards's editorial. Nonetheless, Richards's repeated criticisms of the project on this day supplied a moment of quiet tension in the room—expressed through hushed *tsk-tsks* and sheepish glances. I initially attributed this tension to the fact that Beaula (who I recognized from the documentary) was present as the geneticist delineated the two possible, but possibly conflicting, accounts of her ancestry. However, at the conclusion of Richards's talk, it became clear that Beaula was not the only person in the room who might be experiencing genealogical disorientation.

During the question-and-answer period, a self-described "Grenadian-born British" woman, who appeared to be in her late fifties and who I will call June, announced that in addition to herself, "twelve or fifteen women came up Bubi in the [Motherland] study"; she concluded that the test "does not give a complete blueprint of who I am." Here June established herself as the principal interpreter of her genetic genealogy test. Her use of the word "blueprint" with the qualifier "complete" indicates her belief that the genetic test offered only a partial account of her identity. June's knowledge of the "twelve or fifteen" other "Bubis" in the study also undermined her sense of individuality—her desire for "a complete blueprint of who *I* am," as she put it. The genetic findings did not seem to satisfy her criteria for either genetic or social exclusivity.

Many who attended the Motherland gathering stayed after Richards's presentation to ask additional questions and socialize over tea and cookies. After hearing June voice her concerns, I was curious about Beaula's opinion of the second possible interpretation of her genetic genealogy test. Did she, like June, question the test's reliability because of the presence of the "rare" Bubi marker in several other Motherland study participants? Which result, if any, did she accept after learning that her ancestry might also be traced to southeastern Africa? On what basis did she decide between the alternative accounts of her maternal lineage? I introduced myself to Beaula and we began a conversation about her experience as a participant in the Motherland project. Before she could fully respond to my queries, a man who had been sitting next to her at the meeting joined us. She introduced him to me: "This is my brother, Juan. He doesn't speak English. He speaks Spanish [an official language of Equatorial Guinea]." "Your biological brother?" I asked. "My brother from Equatorial Guinea," she responded. From this point on, the discussion continued between the three of us, with me alternating between elementary Spanish with Juan, and English with Beaula (who spoke even less Spanish than I), but drifted from the topic of Beaula's genetic ancestry tracing, to the purpose of my visit to London and Juan's impressions of England. As a consequence, I was not able to inquire further about Beaula's brief, but suggestive, statement of affiliation with Juan on that day. But several days later I exchanged e-mails with both of them.

When posing the question, "Your biological brother?" I wanted to know how Beaula defined her relationship with Juan. Although she is depicted in *Motherland* as living an Afrocentric lifestyle in Bristol, England, her classification of Juan suggested to me a relationship that was more significant than the vernacular term "brother" used by some blacks to refer to others of African descent. Moreover, in an e-mail, Juan informed me that "Beaula esta ayudando a encontrar a mi madre [Beaula is helping me find my mother]"—a mother with whom he had lost contact many years ago, and who may have migrated from Equatorial Guinea to Europe. This statement suggested that Beaula and Juan did not share a mother, nor was it likely that they were members of the same nuclear family. If Beaula did not regard Juan as a relation in either of these two senses, perhaps "my brother from Equatorial Guinea" described what Catherine Nash terms "genetic kinship," affiliations fashioned from the facts of DNA analysis, the particulars of which are both unspecified and ahistorical. Genealogists also use the phrase "DNA cousin" to characterize persons who might share a set of genetic markers in a genealogical analysis, but whose relationship to one another remains unspecified. In providing associations that are underspecified, genetic genealogy tracing presents consumers with the paradox of *imprecise pedigree.* Root-seekers' awareness of this paradox is indicated by their use of ostensibly redundant phrases such as "DNA cousins" and "genetic kin."[28] These composite descriptors, of course, acknowledge DNA analysis as the medium of affiliation. However, because the words "cousin" and "kin" are already commonly understood to connote "biogenetic ties," the placement of the adjectives "DNA" and "genetic" before these words should be unnecessary.[29] Thus the circulation of these phrases seems to suggest that the associations supplied through genetic genealogy are qualified and, therefore, must be rhetorically set apart from "natural" kinship; or, in other words, that the results of genetic genealogy testing are categorical but imprecise.

What was more certain was that there was something that linked Beaula to Juan and, moreover, made her feel obligated to assist him with his own familial search. Sponsorship of an orphanage and school on Bioko Island and the cultivation of a growing number of Equatoguinean acquaintances, in both Africa and Europe, also indicated Beaula's chosen affinity and showed DNA kinship to be ties that can bind—no matter their imprecision. Her assertion of a familial and ethnic tie to Juan confirmed that, whatever her feelings about the genetic link to Mozambique, she was committed to the ancestry designation she received as a participant in the Motherland study. Similar to Pat's experience, Beaula's genetic genealogy result gained traction through social ties.

For half a century, *Roots* has been an enduring parable of racial slavery and its aftermath: our public memory of a past many do not wish to acknowledge and a touchstone for present-day matters concerning identity, belonging, and community. Haley's model heralded the democratization of genealogy—once an elite pursuit—to broader segments of society. The introduction of genetic ancestry testing pushed this diffusion further, outsourcing genealogy to DNA analysts, but leaving the root-seeker in place as the agent of her genealogical aspirations. In this process, genetic ancestry testing has extended the influence of the *Roots* narrative and its power to inspire forms of reconciliation—personal and familial, diasporic and interracial, transnational and political. At base, the pursuit of African ancestry is an exploration into one's specific family history. Yet root-seeking may also be a journey of orienting satisfaction, disorienting discovery, or historical reckoning. The beginning of a process of reconciliation rather than its culmination, the genetic pursuit of African ancestry begets other pursuits, including the creation of alternative social worlds with reimagined kinship arrangements and affiliations, and hopes for what all of this might become.

FIVE

Roots Revelations

> Washington has variously referred to himself as an American, an African American, a "DNA Sierra Leonean," and an "American–Sierra Leonean," or simply as a "Sierra Leonean."

In the last decade, a spate of genealogy-themed, unscripted (or "reality") television shows, including Henry Louis Gates Jr.'s *African American Lives*, have highlighted the ease and immediacy with which the roots endeavor can be undertaken, be it carried out for a root-seeker by another individual (e.g., a certified genealogist) or a company (such as Gates's African DNA, which sells traditional and genetic genealogy services). On this novel family-history landscape, the apex of the roots journey is "the reveal," a familiar concept in reality television.[1] In this case, new or surprising information, often based upon genetic test results, is presented to a subject who expresses astonishment or elation or both, before an audience. Thus, in the post-Haley era, the practice of root-seeking might be said to involve not simply the reconstruction of a familial narrative, but also one's response to this genealogical account in the presence of an audience. The public reveal reminds us is that the work of reconciliation and repair that genetic ancestry is used to accomplish is always also about a larger group, be it an audience or a community. For the descendants of slaves, this form of public witness may also be a political occasion—a demand that others make note of the sobering historical dynamics out of which some American family trees grew.

The reveal is an essential element of genealogy-themed television shows such as *Motherland: A Genetic Journey* (2003), *Motherland: Moving On* (2006), and *Who Do You Think You Are?* (2004–) on the BBC, and in the United States, celebrity-driven shows such as *African American Lives* (2006), *Oprah's Roots* (2007), *African American Lives II* (2008), *Faces of America* (2010), *Finding Your Roots* (2012–), all on PBS, and NBC's *Who Do You Think You Are?* (2010–).[2] Media scholar June Deery writes that the reveal functions "both to uncover and to display . . . to a dual audience of subject and TV viewers."[3] With televised genealogy shows, furthermore, what is uncovered or displayed—most often to a root-seeker via a host—is information about a notable predecessor, a significant historical event, or unexpected affiliations. The poignancy of these reveals is manifested by root-seekers as heightened emotion or with the flat affect of shock.

For example, in *African American Lives*, a show that featured the genealogy of prominent blacks, genetic genealogy results destabilized long and dearly held ideas about ancestry and identity. Social scientist Sara Lawrence-Lightfoot, who self-identifies as African American and American Indian, was stunned when host Gates disclosed that racial composite testing suggested she had "no Native American" ancestry whatsoever. Astronaut Mae Jemison, the first black woman to travel into space, on the other hand, is pleasantly surprised to learn that her composite includes an inference of 13 percent "East Asian" ancestry. Similarly, during a striking moment in *African American Lives II*, the comedian Chris Rock is brought to the brink of tears when he learns from Gates that a previously unknown forbearer bootstrapped his way up from slavery to two stints in the South Carolina legislature.[4] Conventional and genetic tracing also yielded unanticipated results in 2010's *Faces of America*. Reveals, coupled with the celebrities' raw reactions to the information conveyed, often deliver moments of high drama and genuine emotion. Amid discussion of Malcolm Gladwell's roots, for example, Gates discloses that the best-selling author's Jamaican maternal ancestor, a free woman of color, owned slaves of African descent.

In the post-Haley era, in which the archival labor of genealogy can be at a remove from the root-seeker, genealogists may accordingly take on a new role: no longer solely family-history archaeologists engaged in the lonely pursuit of excavating vital records and census documents, they can become performers whose job it is to react to genealogical information that is revealed to them. Perhaps unsurprisingly, then, less prominent root-seekers than those featured on televised genealogy programs have taken to social media sites to record, perform, and broadcast their reveals, and to disseminate their reflections on the genetic-genealogy-testing experience.

A key to African Ancestry's success has been Kittles and Paige's use of and participation in the public ancestry reveals of prominent blacks as a way to promote interest in its genetic genealogy service. Years before *African American Lives* and *Finding Your Roots* became standard television fare, Kittles and Paige were sharing their MatriClan and PatriClan test kits with newscasters, singers, actors, and other African American notables. Although the genetic genealogy reveal has now become commonplace, when I met with Paige at African Ancestry's headquarters in Washington, DC, she noted that her company was the first to introduce it. African Ancestry was the first genetic-ancestry-testing company to highlight celebrity root-seekers, and the public reveal was a strategy developed to draw attention to its products. "Absolutely, we pioneered this," Paige asserted. The new company had limited marketing resources, and celebrity endorsement was a strategy developed to draw attention to African Ancestry's services—and also to the social import of the work. Its first celebrity client was actor LeVar Burton in 2003. "People were calling us a twenty-first-century *Roots*. . . . What better way to represent the twenty-first-century *Roots* than to tell Kunta Kinte where he's from," Paige said of the actor who first came to acclaim as the protagonist of the television miniseries based on Haley's book (Burton was associated with the Hausa people of Nigeria). African Ancestry's early celebrity reveals included US Representative Diane Watson of California, actress Vanessa Williams, and actor Isaiah Washington in 2005. For Washington, the experience would prove to be especially transformative.

BIOLOGICAL PARENTAGE

After service in the US Air Force, Washington attended the historically black Howard University, where he studied theater. He went on to gain fame as a serious actor with memorable roles in several movies by the director Spike Lee and, beginning in 2005, on the television series *Grey's Anatomy*.

In 2006 Washington was accused of making a homophobic remark against a fellow actor on *Grey's Anatomy* and was summarily released from the show in 2007. Although he later apologized, Washington was something of a pariah in Hollywood for a period. It was during this time, when the pace of his life slowed, that his abiding interests in African culture and history—which up to then had been manifest primarily in Washington's preference for Afrocentric dress and Pan-African politics—became more pronounced. In December 2006, as the Hollywood controversy was just starting to brew, for example, he participated in a White House Summit on malaria, serving as the master of ceremonies for a symposium hosted by President George Bush and First Lady Laura Bush.

A year earlier, at the Pan African Film Festival, Washington was presented with the Canada Lee Award, named for the pioneering and multitalented late black actor and performer, who was red-baited out of his Hollywood career, dying in poverty and without the recognition he deserved for his many accomplishments in theater and music. As the Lee awardee, Washington received MatriClan and PatriClan tests from African Ancestry and sent in his DNA several months before the festival took place in anticipation of a reveal during the event.

Washington initially had trepidations about genetic genealogy and worried in particular about how else his DNA sample might be used. He had "concerns about cloning and having my DNA out there somewhere to possibly fall into the wrong hands." He called Paige, who put him at ease, telling him that "African Ancestry is a privately owned company with no attachments to any forensic or government institution."[5]

Born in Houston, Texas, to Isaiah Washington III and Faye Marie McKee, Washington has stated that his interest in genetic genealogy testing issued from the feeling that he was an adoptive child. "There was still this pull of me wanting to know my biological parents, which would obviously be the people of the continent of Africa. I just wasn't satisfied with the explanation that we were from down south. I wasn't satisfied with the explanation of the history of slavery. . . . I wasn't satisfied that our names were changed and now we have to move on."[6]

On the evening that he received his genetic ancestry test results—his "biological parentage," in his words—he stood on the stage of the Magic Johnson Theater in Los Angeles. Diane Watson and Vanessa

Williams had received their genetic genealogy results and Washington was up next.

> I stood tightly gripping the African staff that was the PAFF Canada Lee Award I had just received. Dr. Kittles approached me holding a reddish brown–colored folder. The room . . . seemed to go still. . . . I began to feel dizzy, and my legs felt weak; still, I refused to succumb. I felt transformed and complete at that moment. . . . I heard him say, "Isaiah, your results show that you share ancestry with the Mende and Temne peoples of Sierra Leone." . . . I felt reborn that night. No longer did I need cowrie shells hanging from my locks, African jewelry, African dance classes, or African drumming circles. . . . All the external things that I thought I needed to connect me to Africa were now unnecessary. Africa had been inside of me all along.

The company's analysis inferred that Washington shared Y-chromosome DNA—which follows patrilineal succession—with one or several Mbundu individuals in contemporary Angola. Using mitochondrial DNA analysis that traced Washington's matrilineage, African Ancestry inferred that the actor shared some similarity to a Mende or Temne person or persons living in Sierra Leone.[7]

Genealogy is as much about the needs and desires of the living as they are portals to the past. So I was unsurprised when Washington, like many other genetic genealogists I interviewed, exercised choice in selecting which of the inferred associations was most significant to him. "Women come first," he said to me, by way of explanation—equal parts chivalrous and paternalistic—for why his post-DNA test identity and activities have been oriented toward Sierra Leone rather than Angola.[8] Since receiving his genetic genealogy results, Washington has variously referred to himself as an American, an African American, a "DNA Sierra Leonean," and an "American–Sierra Leonean," or simply as a "Sierra Leonean."

BORN AGAIN

The desired effect of public reveals may be accomplished through the creation of videos posted on YouTube and other social media platforms that allow response from others. Jasmyne Cannick, a Los Angeles–based cultural critic and political commentator, is one of many who made use of social media to stage her own reveal. She came to African Ancestry for her testing via Isaiah Washington, whom she knew through social circles in Los Angeles.

Personal reveals provide not only a way for root-seekers to circulate their genetic test results, but also an audience with whom to share their experiences and, potentially, develop ties. Viewers' reactions run the gamut from positive to negative. Audience members claiming ties to the ethnic groups or countries to which a root-seeker has been associated by a testing service, for example, may enthusiastically receive (and thus authenticate) a broadcaster's results. At the same time, some in the audience may reflect skepticism about genetic ancestry testing and, implicitly, about the presuppositions about kinship and community that undergird it.

Roots revelations customarily involve the public disclosure of one's previously unknown genetic affiliation (i.e., racial-composite, ethnic-lineage, or spatiotemporal analysis). Although some genealogists reveal their findings only after reviewing them privately, most of this set choose to videotape the climactic moment when their test results are opened. Whether or not the results are revealed "live" on camera, the root-seekers typically share their results with theatrical flair.

Self-described "African" audience members may challenge root-seekers' claims of affiliation—ancestry certificates and performative reveals notwithstanding. In the comments that accompanied Cannick's revelation video, a debate ensued about African American root-seeking. A viewer using the Nguni name "bongiwe" expressed deep skepticism of black Americans' genealogical aspirations and was dismissive of the measures to which they were willing to go in pursuit of African ancestry: "As an African I find it sad and forever tragic the legacy of slavery and the impact it has left on Black Americans. Constantly in search of an identity. There are endless salesmen and business [*sic*] willing to sell you an identity if you are desperate enough and willing to buy it."[9]

Another commenter argued that affiliation with Africa had to be established through action, rather than inferred through DNA analysis. "Shoshaloza1," whose user name refers to a genre of South African call-

and-response folksongs, retorted: "African American? No, no, no, no, no, no. It's American black! The only Africans in America are the ones who were born in Africa! If they consider themselves so African, they should come to Africa and use their talents here to strengthen the reputation of Africa! Instead, they . . . strengthen the reputation of America! . . . Now it is time for American blacks to come back to help Africa. Otherwise, they shouldn't even call themselves African!" Somewhat in agreement, "9revolta" contended that an "African American is someone who has come from Africa, and has gained citizenship in America."

On the surface, the discussion that transpired in reaction to this video appeared to be a dispute over nomenclature. Yet it was also a contest over the stakes of black Americans' claims to African identity. For Shoshaloza1, in particular, there was recognition that genetic genealogy testing, for all its feel-good potential for US blacks, was an asymmetrical exchange, offering little material gain for this viewer and other self-declared "Africans" in the social network.

Skepticism about genetic genealogy was sometimes couched in a discourse of value. The persistence of the question "How much did it cost?" should also be regarded as a question of value beyond strictly economic concerns. "How much does it cost?" might be understood to also mean "Is it worth it?" The latter question goes beyond a cost-benefit analysis to consider the emotional or moral value of the pursuit of roots through consumption. These questions also suggest that acquiescence to a genetic view of kinship brings with it both benefits and sacrifices (that is, "costs" of a nonmonetary type). This aspect of genetic genealogy was on display during a poignant interlude during Cannick's roots revelation. Immediately following her reveal, she is filmed excitedly calling her grandmother to share her results. In response to the news that their family may have West African ancestry, Cannick's grandmother intones, "We're from South Carolina." The granddaughter root-seeker counters, "Our family is from Equatorial Guinea and Cameroon . . . and not South Carolina! [She faces her friend who is videotaping her reveal to share her grandmother's comments.] Grandma is saying that she has 'been fine for 87 years not knowing where she comes from and is saying that . . . those genetic genealogy test companies can't tell people anything."[10]

With notable exasperation and in the hopes that documentation of the results might quell her grandmother's suspicions, Cannick says, "I'll show you the certificate when I come over tomorrow."

The "Certificate of Ancestry" that African Ancestry gives to its clients as part of the results package is featured prominently in several roots revelations. This document, signed by the company's chief science officer, Rick Kittles, illustrates the test subject's genetic ancestry designation in attractive detail. African Ancestry's "certification" is of symbolic importance to root-seekers like Cannick, who declared that she felt "complete" after receiving certified documentation that affiliated her with both the Bubi people in Bioko Island, Equatorial Guinea, and the Tikar tribe and Fulani people of Cameroon. She regarded her certificate not only as a mark of identity, but as analogous to an official vital record. During her roots revelation, she exclaimed, "*I have a new birth certificate!* . . . Now, when people ask me where I'm from, I can say, '[Do you mean] pre- or post- Middle Passage?' "[11]

Reacting to Cannick's roots revelation, YouTube viewer "xbkzfineztx" wrote, "Congratulations on finally knowing YOU. I did [my genetic ancestry testing] recently, which prompted me to look at [your video] . . . I want to take one more test by a different company, which will be African Ancestry. *I want my birth certificate too, shoot!*"

THE "SARA"

Isaiah Washington, whose life's aims after uncovering his genetic ancestry now include highlighting the importance of Sierra Leone to the descendants of slaves in the American South, was on hand on an overcast morning in February 2009 when I arrived in Charleston, South Carolina, to attend "a ceremony of remembrance" for the ancestors dispersed or lost by the transatlantic slave trade. The majority of the people gathered that day laid claim to Sierra Leone in some manner. The ceremony was organized by the James Madison University–based anthropologist and historian Joseph Opala and other participants of "homecoming" pilgrimages from the United States to Sierra Leone that had taken place in 1989, and again in 1995 and 2005, as a way to connect the Gullah and Geechee communities of South Carolina to the history and culture of the West African region once known as the "Rice

Coast." (Rice was king in the eighteenth century on plantations in South Carolina; plantation owners eagerly sought out slave laborers from this West African region who were experienced with its cultivation.)

During these three pilgrimages, Sierra Leonean elders performed ceremonial rites in which they summoned the "common ancestors" of the Americans and the Africans to "bless their homecomings and bring their broken family back together." On this February day, past homecoming participants gathered on the shore of the Ashley River, where slave ships had disembarked in the eighteenth century, to perform the same ceremony on US soil. The morning's spiritual leader for the "ceremony of remembrance" was Amadu Massally, a Sierra Leonean immigrant who had traveled from his home in Texas to officiate at the ceremony. (Massally would return to Sierra Leone later that year after living in the United States for more than twenty-five years.)

Thomalind Martine Polite stood beside Washington, the self-described "DNA Sierra Leonean," on the river's bank. On this morning, they stood together as "kin." Polite traces her roots back to Sierra Leone, through conventional archival evidence, to a little girl named Priscilla who was purchased and transmitted from Bunce Island, a British slave fort near what is now Freetown, Sierra Leone, to a coastal region of South Carolina.

In a best-selling 1998 book, *Slaves in the Family*, Edward Ball traces the roots of his white wealthy southern family and, with admirable genealogical honesty, includes in his ancestral portrait the enslaved men and women his family owned.[12] In contrast with the 2015 controversy in which the actor Ben Affleck successfully lobbied to have his family's slave-owning past hidden on *Finding Your Roots*, the dauntless Ball shined a bright light on the bondage that lies not only at the center of his family but also at the center of American history and society.

At the height of racial slavery, Ball's family had the distinction of being the most successful plantation owners in South Carolina. The Balls had close to two dozen plantations, owned thousands of slaves, and kept detailed and extensive records about the people who labored for them without pay on their rice plantations. Their twentieth-century descendants would use these records intended for commerce and capital accumulation to draw the family tree of the enslaved ten-year-old African girl Priscilla. Ball retraced the family line for seven generations, from Priscilla to Thomas Martin. For his part, Opala located the uncharacteristically detailed cargo manifest of the slave ship that was Priscilla's Middle Passage. The slave manifest and genealogical research together provide the hard evidence of the Polites' African origins—information that is the envy of many slave descendants.

When Thomalind Martin Polite, Thomas's daughter, traveled to Sierra Leone in 2005, she was believed to carry Priscilla's spirit with her. The homecoming journey was said to allow the spirit of the little girl, taken from her home hundreds of years ago, to come to rest in the land of her birth. A ritual was performed in which African Americans and Africans called upon their "common ancestors" as a way to restore their broken families. During this trip, Thomalind met with Sierra Leone's president, Ahmad Tejan Kabbah, and at a waterside ceremony not unlike the one taking place in Charleston, she received an African name.[13]

Massally, Washington, and Polite formed a troika at the center of the intimate group of about twenty-five people that assembled at a bend of the Ashley River. A slave ship from Bunce Island was known to have docked there in 1760, holding a public auction of enslaved Africans. Today the site, known as the Ashley Ferry Landing, is tucked behind a subdivision of unassuming late-model suburban homes.

A few members of the Gullah community—the descendants living in the "low country" or the low-lying coastal regions, were present, including Emory Campbell, who for decades worked to highlight the African linguistic and cultural retentions between his community and Sierra Leoneans. Campbell was a leader of the "Gullah Homecoming" to Bunce Island in 1998. Also in attendance were one or two reporters and a few scholars.

While solemnly singing, the small group passed around a white floral wreath until it came back to Washington, who gently placed it into the river. Massally performed an ancestral sacrifice (or "sara"). He asked the ancestors to join the group, welcomed their arrival, and offered them comfort by providing them with a little of their homeland represented by sand from Bunce Island and rocks from Sierra Leone, which were dispersed into the Ashley. This material symbolized the return of the lost West African homeland to the Sierra Leoneans taken captive generations ago.

This ceremony of remembrance included a group of about eight people who also referred to themselves as "DNA Sierra Leoneans." Several were from South Carolina and many, like Washington, had employed African Ancestry's services in an effort to access even a measure of the family knowledge possessed by Priscilla's descendants. Also in the group were four black women dressed as bondswomen, cultural workers at the nearby Magnolia Plantation, where later that morning Washington would preside over the dedication of small cabins built between 1850 and 1900 in which enslaved people and, later, sharecroppers and paid laborers resided from the era of slavery well into the late twentieth century. The cabins were being opened for heritage tourism.

About an hour later and just a few miles down the road, Washington stood at a podium under a white tent on the grounds of Magnolia Plantation, ready to officiate the dedication of "The Magnolia Plantation Slave Cabin Project." The four cabins had been refurbished with a $100,000 grant from the Annenberg Foundation for their educational value in relaying blacks' experiences on southern plantations. The women in period dress who had taken part in the ceremony at the river took up their posts as historical interpreters, sharing information about the lives and lifeways of the black people who resided there. Washington was joined as a special guest by Joseph P. Riley Jr., the mayor of Charleston, South Carolina; Campbell, in his capacity as the chairman of the Gullah Geechee Cultural Heritage Corridor Commission, and Opala.

Winslow Hastie, a thirty-something fourteenth-generation descendant of the white family who owned the plantation-cum-tourist site, had introduced Washington, who spoke facing one of the cabins. Explaining his presence and his connection to South Carolina and Sierra Leone, he proclaimed to the assembled crowd, "DNA has memory." This statement appears in a brochure for the Gondobay Manga Foundation, a philanthropy established to provide health, education, and economic development services to Sierra Leone by Washington, who became dedicated to "connecting the dots" to the country after receiving his test results. Like many root-seekers, he soon traveled to the country with which he was matched, making his first-ever trip to Sierra Leone in 2006. The highlight of the trip for the actor was his induction into a Mende community and his renaming as "Chief Gondobay Manga." Washington's African name would be used as the name of his foundation, which "advocates cooperative planning to achieve positive, timely improvements in the lives of the people of West Africa."[14] The fundraising efforts of the Gondobay Manga Foundation have to date facilitated the commencement or completion of three significant undertakings.

According to the foundation, an elementary school—the Chief Foday Golia Memorial School, named for a late community leader—was completed in 2008 in the village of Njala Kendema and serves more than 150 students. Renovation of a municipal hospital in Bo, Sierra Leone's second-largest city, is underway. Washington has also been working to help create the infrastructure for a water-filtration system that would provide clean water to the nation's villages. The foundation is also involved with fundraising for the preservation of Bunce Island, the British slave castle, or trading site, that was a point of disembarkment for thousands upon thousands of Africans who would be rendered bonded chattel when they reached the shores of Georgia, Florida, and the Ashley River in South Carolina.

Unsurprisingly for someone who's been in front of a camera most of his life, Washington brought a film crew with him to record his "family" reunion on the African continent. The resulting program, *Isaiah Washington's Passport to Sierra Leone*, was broadcast on the Africa Channel in 2010. Extending the reveal, it followed the actor for four years, from his first trip to the country to the visit that culminates with his gaining dual citizenship there based on his genetic genealogy test.

SIX

Acts of Reparation

> They owe us more than they could ever pay. . . .
> They stole us from our mothers and fathers and
> took away our names from us.

Every generation of African Americans has its reparations struggle. Collective memory of chattel slavery grows dimmer as the years since Emancipation pass. But the drumbeat for restitution—to amend the intergenerational devastation wrought by racialized human bondage—persists, sounding with renewed intensity in each decade, despite historical amnesia. By the end of the twentieth century, the double helix had become a part of this call for reparations.

The genetics zeitgeist is sweeping. Our DNA hopes are more boundless than we often fully apprehend or dare to admit. In 2004, geneticist Kittles, whose ambitions for ancestry testing have always included racial justice and social transformation, found his company's techniques engaged in an effort to obtain slavery reparations. African Ancestry's matrilineal and patrilineal DNA analyses were engaged as twenty-first-century tools that might offer new leverage in the long-waged battle over the repayment of a debt now four centuries overdue.

This novel reparations strategy was born out of a collaboration between Kittles's African Ancestry company and an African American lawyer named Deadria Farmer-Paellmann. The two met as graduate students at George Washington University in the late 1990s. Kittles was pursuing a doctoral degree in molecular biology; Farmer-Paellmann was working toward a master's degree in political management and lobbying. Soon to be known as the "Rosa Parks of the reparations litigation movement," Farmer-Paellmann would conceive a legal plan for restitution for slave descendants that highlighted the connection between inheritance and genealogy and employed DNA to draw these links. The introduction of plaintiffs' genetic-ancestry-testing results as evidence in *Farmer-Paellmann v. FleetBoston* was a strategy that became necessary as the case winded its way from lower to higher courts.

Yet this maneuver was also consistent with the growing utility of genetics across contemporary society—the social life of DNA. Indeed, the use of DNA evidence in this case had two prior, necessary touchstones: The first, discussed previously, was the use of genetics, at the end of the twentieth century, in an array of efforts to rectify past injury to social groups and communities. Forensic inquiries in post-junta Argentina, post-apartheid South Africa, and elsewhere married genetic technology and justice claims and were carried out under the banner of international human rights frameworks that would be adopted by reparations activists.

In the United States, there was also growing awareness of the successes of the Innocence Project, a nonprofit legal-advocacy organization established in 1992 that uses DNA evidence to liberate wrongfully convicted persons, including many African Americans. These high-profile exonerations played a role in alleviating blacks' apprehension about genetic science. As Pat, the genealogist introduced in an earlier chapter, who had worked as an analyst in a crime lab, would say to me: "I'm not question[ing] about DNA. . . . Given my experiences, there is no reason to doubt the technology." Pat articulated what many African Americans expressed to me about the promise of DNA to set black people free—literally and figuratively. Lending a heroic cast to genetics, the Innocence Project offered the double helix as a winding path toward justice and not merely an invidious "back door to eugenics." But concerns about abuses of the new technology are warranted. Discriminatory law enforcement practices in the United States have yielded the largest and most disproportionately black and brown prison population the world has ever known. DNA databases have swelled as a result of a wider range of contact with police, who can collect DNA when making an arrest, sometimes before charging a person with a crime. Law enforcement can also conduct "familial searches" that target not only criminals but their family members. Such practices potentially threaten the civil liberties of innocent people.

Second, the sociopolitical backdrop against which the decoding of the human genome unfolded also played a role in the use of genetics in reparations politics. As we have seen, the Human Genome Project (HGP), completed in 2003, was a technological watershed that ushered in DNA's eventful social life. The project carried mixed ideological messages about the simultaneous irrelevance and salience of race. The human genome comprised the DNA contributions of women and men of different backgrounds, suggesting that our humanity is fundamentally shared and, indeed, interchangeable. Yet, on the other hand, the analogous Human Genome Diversity Project (HGDP) was premised on the belief that there were genetically isolated and distinct racial and ethnic groups to which researchers should urgently attend. Although these distinct projects, the HGP and the HGDP, were not formally linked, they embodied two trains of thought about human difference that would find their confluence in debates about the significance of race after the genome was decoded.

Irrelevant or salient? Scholars across the sciences and social sciences continue to debate the issue. What is certain is that while race may be spoken of in the language of biology, it is fundamentally a political category. It is a way to sort human communities in such a way as to justify social inequality; this sorting is neither natural nor inevitable. What this post-genomic moment did accomplish was the foregrounding of genetics as a lexicon for racial *politics*. DNA analysis was perceived as a new language of social justice, as the moral authority of "inadequate and besieged civil rights discourses" waned. In a climate in which many wished to believe that racial inequality no longer existed, why not then try to use genetics to bolster justice claims?[1]

What began in the late 1990s in Buenos Aires and Johannesburg as forensic projects of identity recovery for the cause of human rights was, by the early 2000s, an endeavor that put genetic identification into the service of a campaign for racial justice, of which reparations for slavery was a facet. The slavery-reparations legal case was, among other things, an attempt to articulate the depths and persistence of racial inequality at a moment when it was said to be nonexistent. Prospects for slavery reparations have always been tenuous, but this was perhaps never so true as at a historical juncture when bold proclamations were being made about race being a non-factor in American life. New strategies were needed.

THE DEBT AND A BROKEN PROMISE

The 2002 class-action suit for reparations for slavery extended a centuries-long struggle. In the immediate antebellum period, the federal government acknowledged that a debt was owed to the newly emancipated blacks who had worked in bondage and without pay. "The debt"—to use reparations activist Randall Robinson's succinct phrasing—begins to accrue with the arrival of the first enslaved Africans at Jamestown, Virginia, in the seventeenth century. A plan for reparations was put in place by the state in 1865. Today's protracted struggle for slavery reparations was inaugurated with the callous breaching of this social contract. At the close of the Civil War, General William Sherman issued "Special Field Order No. 15," a directive that set aside a large swath of land—confiscated by the federal government from Confederate soldiers—including portions of the southern states of Florida, Georgia, and South Carolina. This land was to be divided into forty-acre segments and redistributed to newly emancipated slaves, some of whom had fought in the Union army. By the summer of 1865 "some 40,000 freedmen had been settled on 400,000 acres of 'Sherman land,'" and plans were in place to distribute another 450,000 acres of land as restitution. In addition to land, each formerly enslaved family was leased or loaned a mule. Accordingly, the refrain that emancipated men and women were promised reparations in the form of "forty acres and a mule" for the discounting of their humanity and their loss of wages would soon and for long after travel in African American communities.[2]

Within a few months, African Americans' march toward recompense came to an abrupt end. In September 1865, following the assassination of President Abraham Lincoln, newly elevated president Andrew Jackson overrode Sherman's order and commanded that the land allotted to freed blacks be returned to Confederate soldiers, who would be pardoned in exchange for pledging their allegiance to the United States. This White House decree, called Howard's Circular 15, was drafted by and named for Freedmen's Bureau commissioner Oliver Otis Howard, to whom would fall the task of seeing that this prejudicial policy reversal was carried out. Howard's change of duties was a bitterly ironic development, because the Freedmen's Bureau was now asked to implement the disenfranchisement of the very black Americans it had been created to aid and support.

African Americans challenged this breach of contract from the start. Historian Eric Foner recounts the rueful scene at a meeting in South Carolina where Howard was dispatched to explain the policy change to formerly enslaved men and women, who were recent recipients of forty-acre parcels of land:

> When [Howard] rose to speak to more than 2,000 blacks gathered at a local church, "dissatisfaction and sorrow were manifested from every part of the assembly." Finally, a "sweet-voiced negro woman" quieted the crowd by leading it in singing the spirituals "Nobody Knows the Trouble I Seen" and "Wandering in the Wilderness of Sorrow and Gloom." When the freedmen fell silent, Howard begged them to "lay aside their bitter feelings, and to become reconciled to their old masters." He was continually interrupted by members of the audience: "No, never," "Can't do it," "Why . . . do you take away our lands?"[3]

As this account conveys, Howard did not only announce the end of the reparations plan, he also recommended social backsliding, encouraging freedmen and freedwomen to be "reconciled to their old masters." This suggestion was quite the opposite of the racial reconciliation that was desperately needed in the wake of a fractious war in a nation built upon human captivity.

Adding insult to injury, Howard mandated that a committee of recently emancipated black men devise the process by which they would be divested of their land and these plots returned to Confederate soldiers. (In cases in which the transfer of land could not be negotiated, African Americans were violently evicted from the land the state allotted to them.) The black men asked to join this committee voiced the deep sense of betrayal felt by their communities in a formal complaint:

> We were promised Homesteads by the government . . . if the government having concluded to befriend its late enemies and to neglect to observe the principles of common faith between its self and us its allies in the war you said was over, now takes away from them all right to the soil they stand upon save such as they can get by again working for *your* late and their *all time* enemies . . . we are left in a more unpleasant condition than our former. . . . You will see that this is not the condition of really freemen.[4]

The government's abandonment of its reparations policy left African Americans in "a more unpleasant condition" of destitute poverty and subjugation to both those who enslaved them and to those with whom they fought in supposed solidarity. This betrayal did not go unchallenged. In subsequent decades, the outrage voiced here would gather intergenerational force, becoming a peculiar bequest passed down to descendants from ancestors. The African American reparations movement was born from this acrid cauldron of disloyalty and disappointment.

A REPARATIONS-MOVEMENT GENEALOGY

Since this time, attempts to compel the United States—a nation literally and figuratively formed through the economy of slavery—to pay its debt to African descendants have continued. The topic of this chapter, *Farmer-Paellmann v. FleetBoston*, a historic slavery-reparations lawsuit against multinational corporations, that in another first would engage genetic ancestry tests as evidence, is but a recent milestone of a much longer journey for racial reconciliation and economic restitution.[5] Twenty-first-century activists see themselves engaged in a struggle to right wrongs dating back to the era of plantation slavery, but this politics now includes contemporary social challenges that stem from the history of racialized human bondage, with both past and present now viewed through a global prism, articulated through the language of international law and the discourse of human rights, and pursued through diverse strategies.

US history is punctuated with instances in which bondsmen and bondswomen and their descendants have endeavored to receive restitution for the forced, unpaid labor of their ancestors and the corollary damage of plantation slavery. These demands have taken many forms, including moral appeals, economic calculations, legislative efforts, and lawsuits. Despite advancing varied strategies toward achieving recompense, over the last 150 years, prospects for the payment of reparations have never been promising. Nevertheless, from the lobbying efforts of a late-nineteenth-century grassroots movement of ex-slaves to the use of DNA analysis by a group of slave-descendant litigants in the early twenty-first century, each campaign has advanced with dogged ingenuity in the face of a recalcitrant nation.

African Americans' antebellum disappointment was soon channeled into organized efforts to seek compensation for the formerly enslaved. The multifaceted Ex-Slave Mutual Relief, Bounty and Pension Association led one such campaign. Established in 1894 in Nashville, Tennessee, by Isaiah Dickerson and Callie House, this chapter-based national organization had two central aims. The first was the practical task of offering emancipated blacks the social welfare support that the Freedmen's Bureau no longer provided after Congress disbanded it in 1872. To this end, the organization's local chapters, which partly functioned as mutual aid societies, collected monthly dues from its membership that were redistributed to members in times of need (e.g., infirmity, disability, for funeral expenses, etc.). Second, working on the national stage, the group contended that formerly enslaved persons were entitled to pensions such as those paid to former Union soldiers. For more than two decades, the Ex-Slave Association would raise public awareness around the pension issue and lobby politicians to support related legislation. Notably, pension proponents included not only this group of formerly bonded persons, but also prominent and often paternalistic whites, such as Dickerson's former employer, newspaperman and Confederate veteran William Vaughn, who had begun to advocate for pensions several years before the formation of the Ex-Slave Association in the hopes that the plan would produce financial stimulus for the impoverished South.[6]

Besides Dickerson, the Ex-Slave Association's most prominent and important leader was House. A former slave, a widow, and a mother, House traveled the United States organizing emancipated blacks and garnering support for her organization and its reparations goals. During her travels, she also petitioned members of Congress to support the passage of HR 11119, a bill that would establish a pension plan for the formerly enslaved. The bill did not move beyond the congressional Committee on Pensions, where it was tabled. Given that congressional leaders never took the activists and their allies seriously, this was unsurprising. Indeed, one government official freely admitted that there was not "the remotest prospect" that the pension bill would be passed into law.

In 1915, with the organization's legislative strategy for reparations effectively stalled, House quietly spearheaded a class-action lawsuit filed by four former slaves against William McAdoo, then secretary of the Treasury. The litigants traced and calculated the economic gains to the state that resulted from the labor of black enslaved workers. The African American plaintiffs contended that their labor yielded cotton that in turn produced tax income for the federal treasury in excess of $68 million during a six-year span from 1862 to 1868. This sum was the least the US government should pay, the litigants argued, for more than a century of expropriated labor. The case went to appeal. In 1916, an appellate court ruled that in naming McAdoo as the suit's defendant, the plaintiffs were effectively suing the US government by proxy. Under the principle of governmental or "sovereign" immunity, however, the federal government could not be sued. As the court ruled, "the United States cannot be made a party to this suit without its consent."[7] The Supreme Court would uphold this decision. Legislative and judicial defeats, and the US government's harassment and intimidation of House—who would be convicted on questionable federal mail-fraud charges the following year, a fate shared with the black nationalist leader Marcus Garvey—effectively marked the end of the Ex-Slave Association.[8] But its work has carried forward.

Estimated to have had a membership of several hundred thousand people at its height, the Ex-Slave Association's multifaceted approach to reparations advocacy, which tacked from grassroots organizing to political lobbying and from economic calculation to courtroom sparring, illustrates the resourcefulness necessary to advance an unpopular cause. While the Ex-Slave Association was not the only organization of its time devoted to securing slavery reparations, its innovative, wide-ranging activism planted the seeds for future movements.

The group's rise and fall also foreshadowed the barriers that subsequent reparations activists would face, callous indifference and dogged harassment among them. More materially, the Ex-Slave Association's path revealed the legal constraints that would daunt any future claims for reparations, including sovereign immunity, the legal doctrine that renders a sovereign state immune from civil suit or criminal prosecution. Indeed, detouring around sovereign immunity was the challenge confronting reparations activists a century later; in *Farmer-Paellmann v. FleetBoston* plaintiffs would pursue restitution from private corporations that benefited from the transatlantic slave trade rather than seek repayment from the state.

By the mid-twentieth century, the cause of reparations was a well-developed focus of African American social justice activism. The efforts of the "Queen Mother" Audley Moore are a case in point. Moore, the Louisiana-born granddaughter of slaves, was an activist and organizer for more than seven decades. She got involved in racial politics in 1919 after attending a lecture by the black nationalist leader Marcus Garvey. "It was Garvey who brought the consciousness to me," Moore would later recall.[9] She joined Garvey's Universal Negro Improvement Association (UNIA) and soon after moved from New Orleans to New York City to work on behalf of the organization. In Harlem, Moore became engaged in a wide range of political activities including reparations activism. "Ever since 1950, I've been on the trail fighting for reparations," she said. In addition to stolen wages and lost wealth, mid-century reparations activists like Moore, who carried out their work three generations removed from bondage, began to shed light on the damage of racial slavery in the present, including the loss of kin and heritage. As Moore explained, "They owe us more than they could ever pay. They stole our language, they stole our culture. They stole us from our mothers and fathers and took away our names from us."[10]

This wider perspective also allowed Moore a broader geopolitical vantage for her reparations activism. Understanding the struggle for black liberation through reparations as a transnational issue amenable to international political remedies, Moore presented petitions to the United Nations in 1957 and 1959 on behalf of slave descendants, charging that "genocide" had been perpetrated against black Americans. Adopted by the UN following World War II, the Convention on the Prevention and Punishment of the Crime of Genocide defined genocide as including practices that were norms of plantation slavery, such as "killing members of the group," "causing serious bodily or mental harm to members of the group," "inflicting on the group conditions of life calculated to bring about its physical destruction in whole or in part," "imposing measures intended to prevent births within the group," and "forcibly transferring children of the group to another group."[11] Moore believed that the transatlantic slave trade comprised genocide and demanded that amends be made to black Americans in the form of land and money. The turn to international statutes reflected lessons learned from the failures of earlier reparations activists using domestic law as well as the ingenuity now characteristic of reparations activism.

Around the same time, Moore read an essay explaining that international law held that "enslaved people [are deemed to be] satisfied with their condition" if they "do not demand recompense after 100 years have passed."[12] This information gave her reparations work a particular urgency. In 1962, a few years shy of the one-hundredth anniversary of the Emancipation Proclamation, Moore demanded reparations for blacks in a White House meeting with John F. Kennedy during which she presented the president with a petition of one million signatures in support of restitution. In the same year, Moore founded the Reparations Committee of Descendants of United States Slaves.[13] The organization was a crucial forerunner in the work of "grassroots education on reparations" in African American communities and would propagate the cause among a new, younger generation of activists.[14]

Indeed, it was Moore who would stir a passion for the reparations cause in Harvard legal scholar Charles Ogletree, during a chance encounter when both were traveling to Tanzania in the 1970s; he would go on to cochair the Reparations Coordinating Committee (RCC) with TransAfrica Forum leader Randall Robinson and Adjoa Aiyetoro, lead counsel for the National Coalition of Blacks for Reparations in

America (N'COBRA).[15] A legal think tank composed of activists, established in 2000, the RCC was devoted to finding a strategy for reparations that could be carried forward in a US courtroom or before an international body. As Ogletree explained, the outcome sought by the group was not financial restitution so much as "a change in America." He elaborated, "We want full recognition and remedy of how slavery stigmatized, raped, murdered and exploited millions of Africans through no fault of their own. . . . The country has never dealt with slavery. It is America's nightmare. A political solution would be the most sensible, but I don't have a lot of faith that's going to happen. So we need to look aggressively at the legal alternative."[16]

At the same time, Deadria Farmer-Paellmann was working on a parallel strategy to that of the RCC, building momentum to bring the class-action suit that bears her name to trial. Among the hallmarks of Queen Mother Moore's championing of reparations was the expansion of it from a domestic issue to a global one in African American politics, by laying claim to international human rights law available after the formation of the UN in 1948. This framework would be central to the RCC strategy and to the *Farmer-Paellmann v. FleetBoston* litigation, categorizing slavery as a "crime against humanity" and, in so doing, appealing to the UN for restitution and reconciliation.

The Republic of New Afrika (RNA), founded in 1968 by brothers Imari Obadele (formerly Richard Henry) and Gaidi Obadele (formerly Milton Henry), also foregrounded the larger geopolitical terrain of reparations activism. Imari Obadele would play a founding role in the important reparations organization N'COBRA.[17]

While Queen Mother Moore and the RNA had an international frame of reference, the context for the "Black Manifesto" was decidedly domestic. James Forman Sr., a former leader of the Student Nonviolent Coordinating Committee, took up the baton in the race for racial recompense in the black power era. On May 4, 1969, he disrupted Sunday services at New York City's historic Riverside Church in order to make an audacious pronouncement demanding reparations for African Americans. (Forman was one of a national network of black activists, working in concert, who interrupted religious services across the United States to deliver the Black Manifesto on this day.)[18] Speaking on behalf of the National Black Economic Conference, whose members had collectively drafted the manifesto in Detroit one month prior, he asked that $500 million in restitution be paid to black Americans by "Christian churches and Jewish synagogues" that were said to be "part and parcel of the system of capitalism" that had contributed to the global exploitation of black people.[19] The Black Manifesto, which would be published in the *New York Review of Books*, reflected the revolutionary rhetoric of the time in seeking reparations in the form of access to education, job training and employment, land in the South to support cooperative farming, and an expanded welfare benefits system. In Ta-Nehisi Coates's widely read 2014 long-form essay, "The Case for Reparations," he maintains that the "compounding moral debts" since slavery—black codes, segregation, economic marginalization—and the nation's lack of an ethical response to this suffering, rather than slavery per se and solely, are what is at issue in the politics of reparations. As the facets of the Black Manifesto make clear, Forman and his collaborators knew this well. Here reparations claims based on the abuses of slavery *and* its aftereffects further evolved to implicate capitalism and racism.

In the final decade of the last century and the first years of the new millennium, reparations politics reached an apex. By the late twentieth century, more than one hundred years after the formal end of slavery in the United States, the reparations movement sought to respond to the historical erasure of the trauma of slavery. Beginning in 1989, US Representative John Conyers introduced bill HR 40—the Commission to Study Reparation Proposals for African Americans Act. This bill, named to resonate with the "forty acres and a mule" that were promised but that went undelivered to newly emancipated African Americans, is a moderate piece of legislation. Its passage would merely mandate the creation of a commission to study slavery and its present-day impacts, and make suggestions about possible remedies.[20] In more than twenty-five attempts, HR 40 has never made it out of the House Judiciary Committee. The lack of support for the Conyers legislation effectively means that the United States "has never authorized an examination of this nation's participation in the enslavement of Africans and the segregation and labor exploitation of their descendants."[21]

Herein lies the rub. Why avoid the mere discussion of reparations?

Coates advises that "the idea of reparations is frightening not simply because we might lack the ability to pay. The idea of reparations threatens something much deeper—America's heritage, history, and standing in the world. . . . Reparations—by which I mean the full acceptance of our collective biography and its consequences—is the price we must pay to see ourselves squarely."[22] In other words, at this moment, acknowledgment and racial reconciliation are what is at stake, and these are perhaps more important than monetary reparations. Historical amnesia is a lynchpin of today's "post-racial" politics—out of sight, out of mind—and bringing the history of racial discrimination into view has become one of the principle jobs of genetic ancestry testing.[23]

In 1994, California was the site of the first slavery reparations lawsuit since *Johnson v. McAdoo* almost one hundred years prior. In the words of the court, in *Cato v. United States* the plaintiffs sought "damages due to them for the enslavement of African Americans and *subsequent discrimination* against them, for an *acknowledgment of discrimination*, and for an *apology*."[24] In keeping with the resourcefulness exercised in 130 years of reparations struggles, the attorneys for *Cato* invoked the legal precedent of a case involving Native Americans. In *Oneida Nation v. State of New York*, the Oneida community won a US Supreme Court case in which they were awarded restitution for "wrongful possession of their lands." The plaintiffs in *Cato* drew a parallel to enslaved men and women as wrongfully possessed human property. However, the statute of limitations and sovereign immunity—the hurdles that doomed prior slavery reparations trials—remained. Thus, the fate of *Cato* was not a surprise. More recently, in *Obadele v. United States*, the Court of Claims dismissed this case seeking slavery reparations based on the model with which restitution was paid to Japanese Americans who were interned during the Second World War. (Notably, the plaintiff in this case was N'COBRA cofounder Imari Obadele.) The plaintiff contended that not to allow African Americans to make use of the Civil Liberties Act of 1988 was unconstitutional because it denied them equal protection. The case was decided against Imari Obadele, with the court noting, however, that the act could apply to other groups. The legal hurdles faced in the *Cato* and *Obadele* decisions would be encountered in any future litigation to obtain reparations.[25]

In 2000 the anti-apartheid activist and Harvard Law School graduate Randall Robinson published the best seller *The Debt: What America Owes to Blacks*. The book is an expansive and impassioned argument for why blacks deserve reparations. Like Michelle Alexander's influential book *The New Jim Crow*, which would be published a decade later, Robinson's book set the agenda for black politics in its moment. He advocated payment in the form of a national trust fund that would finance social, economic, and educational empowerment for African Americans. Importantly, and anticipating how the *Farmer-Paellmann* case would evolve, Robinson argued that the reclamation of lost history was as important as monetary payment. He used the analogy of recent international reconciliation events to press his point, writing that "we make only the claims that other successful group complainants have made in the world. Put simply, we too are *owed*."[26] Like Coates, he stressed the importance of merely opening up a national conversation about the history of racial slavery: "The catharsis occasioned by a full-scale reparations debate . . . could launch us . . . into a surge of black self discovery. . . . We could disinter a buried history . . . and give healing to the whole of our people, to the whole of America."[27] *The Debt*, which can be considered a twenty-first-century manifesto for racial reparations by Robinson, who cochaired the RCC with Ogletree, galvanized interest in financial restitution for black Americans. By 2001, the *New York Times* described the reparations struggle as a movement of "substantial force" that was "gaining steam."[28] The ebb and flow continued.

SEVEN

The Rosa Parks of the Reparations Litigation Movement

> Plaintiffs cannot establish a personal injury sufficient to confer standing by merely alleging some genealogical relationship to African-Americans held in slavery over one-hundred, two-hundred, or three-hundred years ago.

Deadria Farmer-Paellmann was well aware of the fits and starts of reparations politics when she decided to take another tack, moving from legislation and lobbying to the courtroom. She is founder and executive director of the Reparations Study Group, a New York–based nonprofit self-described as an "organization that examines approaches to securing restitution for injuries inflicted upon oppressed people." This group is also involved in uncovering corporate ties to the slave trade. She is an advocate long dedicated to the cause of seeking reparations for the unpaid labor of enslaved Africans.[1]

The African American activist was born in the East New York neighborhood of Brooklyn, New York, and was partly raised in the Bensonhurst community of that borough. As a child, discussions with her grandfather, Willie Capers, left a lasting impression on her. Capers, the grandchild of runaway slaves, often railed against racism, concluding his denunciations with the phrase, "And they still owe us our forty acres and a mule."[2] This embedded in her a sense of "America's broken promise" to the descendants of enslaved men and women and a desire to know more.[3]

Farmer-Paellmann completed a bachelor's degree in political science at Brooklyn College, a City University of New York campus, in

1988.[4] But she found her training oddly apolitical in nature. She then attended the Graduate School for Political Management in Manhattan. Here she focused on lobbying, and this training allowed her to begin to imagine what a persuasive campaign for reparations might look like. Soon after she began working as an administrator in the publicity office for the New York City Health Department, acquiring public relations experience through her interactions with the publicists in that unit.

In 1992 she became involved in the foment surrounding the African Burial Ground in lower Manhattan.[5] It was here that she would first take notice of Rick Kittles's research. It was also at this site that she would begin to link reparations with the legacy and social inheritance of slavery in a deeper way. At the time, Kittles was a doctoral student in genetics at George Washington University whose relationships with Howard University scholars led to his becoming a researcher on the African Burial Ground project. Farmer-Paellmann was tasked with creating a protest at the burial ground site that would draw press attention to the disposition of the graves. "We decided that the media event would be a twenty-six-hour drum vigil," she recalled.[6] At the burial ground, she was the leader of an artistic protest, an immense assembly of people drumming, chanting, and dancing. The activists would succeed in ceasing excavation at the site for a time.

In the course of planning the vigil, she toured the excavation site, and the experience riveted her. During this exploration, she witnessed "archaeologists kneeled over skeletal remains using toothbrushes to push away dirt that had lain undisturbed for two centuries. It was a bit surreal for me to be face-to-face with the actual remains of my African ancestors. . . . Many appeared to have died in pain, their mouths stretched wide open as if they were screaming. The chief archaeologist reassured me that the haunted look resulted from the deteriorations of jawbone muscle that normally holds the mouth shut."[7] This experience forever changed her. As she relayed to the *New York Times*: "They looked like they were screaming. . . . They lived, worked and died in this city, and they never got their reparations—it moved me."[8]

She enrolled at George Washington University in 1994, where she completed her graduate training, earning a master's degree in lobbying and political campaign management. Having completed her training at

George Washington, Farmer-Paellmann set her sights on a law degree. She matriculated at the New England School of Law in 1995, graduating from the Boston-based institution in 1999. She enrolled in law school in order to study legal strategies with which she might bring a reparations suit to trial,[9] embarking on a legal career "with the intent of finding a case for slavery reparations against the federal government."[10] (Farmer-Paellmann also completed some legal training at the Cardozo School of Law at Yeshiva University, where founding Innocence Project attorneys Barry Scheck and Neufeld—who pioneered the use of DNA technologies for another form of black liberation—were on the faculty.)[11] Concurrently, Farmer-Paellmann continued her other activist work, serving as a law clerk for the N'COBRA organization.[12]

At law school, Farmer-Paellmann read a law review article arguing that genealogy could be a possible strategy for bringing a reparations lawsuit forward. The legal scholar Vincene Verdun proposed that the descendants of enslaved men and women could file a suit for unpaid wages against the descendants and agents of former slave owners based on the legal principle of restitution, which holds that ill-begotten assets must be returned to their legal owners and their heirs. Because the federal government would not be named in the case, this strategy had the additional benefit of getting around the sovereign-immunity hurdle that thwarted the efforts of Callie House more than a century prior. This approach also could overcome the statute of limitations obstacle, because such time constraints do not typically apply in restitutions cases.

As a student in the course Race and the Law, taught at the New England School of Law by Robert V. Ward and Judith Greenberg, Farmer-Paellmann prepared

> a case for reparations that required me to research my family roots to link myself to a particular company. To conduct the complicated genealogy research required to trace enslaved ancestors, I referred to the book *Black Genealogy*, by Charles L. Blockson. Blockson suggested that one source of tracing enslaved ancestors was slave life insurance policies. He directs readers to Aetna Incorporated ... and Lloyd's of London. ... In January of 2000 ... I called Aetna to request copies of their slave policies. An enthusiastic archivist

> sent copies of two policies, and a group of circulars from life insurance companies that competed with Aetna in its slave policy business.[13]

Black Genealogy was first published in 1977, the year after Haley's influential ancestry-tracing tome, *Roots*. Blockson's text is a sort of almanac for root-seekers; it contains suggestions for African Americans on how to get started with genealogy, and information about how to access public records relevant to African American history and obtain reproductions of primary documents. Blockson recommends two paths to one's roots—Aetna and Lloyd's of London; these businesses' meticulous records stand in stark contrast to the dearth of other archival information about enslaved persons. For Haley, Lloyd's of London was the key that opened the door to Kunta Kinte. For Farmer-Paellmann, who was aware that a successful reparations claim would require a conclusive genealogical claim, her family history was a vehicle toward her goal. And Aetna provided the documentation that advanced this landmark class-action suit.

MAKING THE CASE

After decades of failed attempts, the possibility for a strategic shift in the legal battle for reparations became apparent in 2000, when at the age of thirty-four and fresh out of law school, Farmer-Paellmann uncovered archival evidence that Aetna, a leading insurance provider, had written policies on the lives of bondsmen and bondswomen for slave owners. Farmer-Paellmann recounts her discovery with Aetna this way: "I obtained copies of two of these slave policies from Aetna's archivist. Seeing documents bearing the words 'Aetna Slave Policy,' I wept and thought to myself, the world needs to know what this multibillion-dollar corporation did. After all, the policies provided slaveholders with security to invest in human chattel with little risk. This practice helped to perpetuate slavery. . . . I contacted Aetna and asked that they apologize and pay restitution. They agreed to do both. By March of 2000, Aetna made an unprecedented apology for their role in slavery; however, they backed down on their promise to pay restitution."[14] This news was widely publicized in the national media. Aetna would soon disclose that in the nineteenth century it had indeed provided insurance to cover slaveholders' losses in the case of the death of their slaves.[15] The revelation would evoke a formal response from the Connecticut-based insurance company in the form of a public apology for its participation in the slave trade.

This discovery would also take on personal relevance for Farmer-Paellmann, who would find evidence that the insurance company held an insurance policy on her great-grandfather, Abel Hines, a slave. This evidence, she argued, was "a discovery that aids in establishing my standing to pursue the action pending in the Chicago Federal District Court."[16]

For Farmer-Paellmann, the quest for reparations was about more than money; it addressed a loss of fundamental humanity. While reparations claimants are criticized in public debates for seeking a redistribution of wealth or, less generously, undeserved handouts, Farmer-Paellmann saw that a deeper form of recompense and recognition was needed. As she would explain to a British publication, "Our injury is that we don't know who we are."[17] Redress and identity are inextricably linked for this reparations activist.

Farmer-Paellmann began her legal studies at a time when a reparations claim by African Americans seemed especially viable.[18] In 1988 Congress voted in support of monetary restitution and a formal apology to Japanese Americans who were interned by the federal government during World War II, and to their families. President George H. W. Bush signed a bill into law—Civil Liberties Act Amendments of 1992—that approved the distribution of reparations to Japanese Americans. Bush carried through to completion a policy initiative that had been initiated by President Jimmy Carter and endorsed by President Ronald Reagan. In 1999 Germany and the United States negotiated reparations in the sum of more than $5 billion for slave laborers who were forced by the Nazi regime to work for well-known manufacturers including Volkswagen and the pharmaceutical company Bayer.[19] These cases formed a basis on which reparations claims for the descendants of enslaved Africans, who later lived under Jim Crow, might be based. The difference, however, was that reparations were being sought by the grandchildren of the enslaved and not by the injured parties themselves, as with the Japanese and German cases.

The publication of *The Debt* that same year, and Farmer-Paellmann's revelation of Aetna's involvement in the slave trade in 2000, lent credence to the call for reparations. When *Farmer-Paellmann v. FleetBoston* was filed in early 2002, it was the outgrowth of recent developments in policy, public discourse, and social activism. Reparations politics can be troubling because they constitute social justice issues as historical calcifications rather than as urgent, contemporary matters. Yet the confluence of events and sentiments that set the stage for this millennial reparations endeavor also embodied a new kind of racial politics that bears notice, and that did not bear only on the past. In historian Martha Biondi's estimation, "Reparations [politics] represent the culmination of a long African American human rights struggle."[20] In this way, these acts of reparation have momentum. They serve as a path to the past and a bridge to the future. And, far from merely symbolic, the intentions of African American reparations activists are deeply material and engaged with both geopolitics and political economy. Given the United States' recalcitrance to publicly acknowledge the enduring *mal* effects of chattel slavery—effects that persist to the present day—to turn the tide of the American narrative on the consequences of racial slavery is no small feat. Such a transformation would be dramatic; it would change the "image of African Americans from victims to creditors."[21]

FROM VICTIM TO CREDITOR

Farmer-Paellmann's class-action suit against FleetBoston and others coalesced from several individual cases in which corporations were sued by the descendants of slaves for the return of lost wages and, consequently, lost wealth.[22] Harvard law professor Ogletree, a member of the Reparations Assessment Group—a "dream team" of attorney-advocates mobilized around redress for black Americans—and a cochair of the Reparations Coordinating Committee, a body that also seeks legal avenues to reparations for slave descendants, praised the originality of Farmer-Paellmann's approach, stating, "The idea of corporate involvement has always been raised in the reparations movement. . . . But I don't think anybody has been as conscientious or as thorough as Deadria. She is the key factor in making these [legal] claims come to life."[23] Farmer-Paellmann's efforts proceeded independently of the Harvard group of lawyers and academics that comprised the effort helmed by Ogletree; the high regard with which the Harvard-trained attorney held the New England School of Law–educated Farmer-Paellmann was a testament to her diligence and courage.[24]

In 2002, in addition to New York State, New Jersey, Texas, Illinois, and Louisiana, slavery reparations suits were filed by eight plaintiffs against eighteen defendants, among them some of the most prominent financial institutions in the United States and the United Kingdom: including the three previously mentioned, these other plaintiffs named Lloyd's of London, New York Life, Union Pacific Railroad, AIG, Lehman Brothers, J. P. Morgan Chase Manhattan Bank, R. J. Reynolds Tobacco, and other banking and industrial giants that had profited as a result of the slave trade. These cases were consolidated into a class-action lawsuit with eight plaintiffs standing in for more than thirty million African Americans who were deemed deserving of restitution from corporate entities.[25] The plaintiffs did not seek a set monetary sum as reparations, but estimated the value of that unpaid labor in excess of $1 trillion.

On March 26, 2002, Farmer-Paellmann's lead attorney, Edward Fagan, filed a historic complaint and demand for a jury trial in a New York federal court on behalf of his client and "all other persons similarly situated," against FleetBoston Financial Corporation, Aetna, and CSX.[26] Fagan was an ideal courtroom advocate. He had successfully obtained a billion-dollar settlement from Swiss banks on behalf of Holocaust victims. He was the attorney who successfully negotiated a $5 billion settlement for slave laborers in Germany during the Second World War, from whose work multinational corporations profited. And Fagan had also been involved in suits against Swiss and US companies that profited during the apartheid era in South Africa. As reporter Robert Trigaux put it in 2002 in the *St. Petersburg Times*, "Farmer-Paellmann's class-action complaint relies on the same idea Fagan used in the Nazi suit. Labor is property. Wrongdoers who take property produced by slaves should give it up."[27] Aetna, which two years prior had declared its "deep regret over any participation at all in [the] deplorable practice" of slavery prompted by Farmer-Paellmann, now offered to establish a scholarship fund for minority students, but was deeply skeptical of the reparations suit and fought it.[28]

The plaintiffs argued that these present-day corporations hold wealth that was obtained through the unpaid labor of slaves, or through ancillary industries that supported the slave trade: Aetna collected premiums from slaveholders for insurance policies written on slaves' lives. Lloyd's of London insured ships that brought slaves from Africa to the Americas. CSX, a railroad company, profited from the transportation of slaves and the use of their unpaid labor to expand its tracks. And Brown Brothers Harriman provided interest-bearing loans to plantation owners enabling them to buy enslaved Africans. The plaintiffs sought a national apology to the descendants of slaves, and financial reparations from these corporations in the form of large trust funds that would be used for social welfare programs to improve housing, education, and healthcare. There was broad interest in this legal strategy, but not universal support for it. Some critics of the class-action suit likened it to blackmail. David Horowitz, the neoconservative author and activist, for example, called it a "shakedown,"[29] while an international reporter pilloried Farmer-Paellmann as a "moral extortionist" in the *New York Times*.[30]

The case was heard in late 2003 by the United States District Court for the Northern District of Illinois. After close to eighteen months of deliberation, in January 2004 Judge Charles R. Norgle, a Reagan appointee, granted the corporate defendants' motion requesting a dismissal of the case. In an opinion of more than seventy pages, Norgle's dismissal was based on several grounds, including the fact that the plaintiffs' claims exceeded the constitutional authority of the court and could only be heard by either the executive or legislative branch, and that the statute of limitations for their claims had long expired.

Significantly, Norgle also found the plaintiffs' case lacking with respect to the legal doctrine of "standing." The judge asserted that the plaintiffs did not demonstrate a precise connection to former slaves, and thus the plaintiffs could not sue for injury as their descendants. Anticipating this particular criticism of their claim, the plaintiffs' suit had included wording that sought to get around the standing doctrine by explaining the significant barriers faced by persons of African descent in the United States trying to trace their genealogical origins. They wanted the judge to understand how and why they had hit a "brick wall," a term I had encountered frequently in African American genealogy circles. As anthropologists Faubion and Hamilton explain, the plaintiffs argued "that their case was hindered by the sheer lack of archival evidence concerning the enslaved ancestors and that the defendants had purposefully withheld and, in some cases, destroyed records and other key documents that might have been used to establish some kind of legally cognizable genealogical relationship between them and their ancestors."[31]

Norgle was not persuaded by this argument. He countered that the companies under scrutiny did not own slaves, writing that the "Plaintiffs who allege that they are descendants of enslaved African-Americans fail to allege that their ancestors were *enslaved* by any of the eighteen specifically named Defendants."[32] In addition to underscoring the fact that the entities being sued were not slaveholders, Norgle argued that even if they had been, the plaintiffs could not "establish a personal injury sufficient to confer standing by *merely alleging* some genealogical relationship to African-Americans held in slavery over one-hundred, two-hundred, or three hundred-years ago."[33] With this line of reasoning, the court contested the plaintiff class's claim of "hereditary or genetic standing" and rejected the assumption of a "familial relationship between the descendant plaintiffs and their enslaved ancestors."[34] Norgle dismissed the case "without prejudice," giving the plaintiffs two weeks to bring the case forward again, submit a new case, or turn to a federal appeals court.

DNA FOR THE PLAINTIFFS

The reparations activists had two responses to the dismissal of their case: one course of action took place outside the court of law and one was rooted firmly within it. First, the activists turned their focus to putting "political and consumer pressure on companies" that benefited from the slave trade.[35] Beginning in 2004, Farmer-Paellmann's Restitution Study Group collaborated with student groups nationwide, including those at Southern Connecticut State University and Clark Atlanta University, to call for a boycott of the financial companies named in the class-action suit that today provided student loans. The "One Student" campaign focused on J. P. Morgan Chase in particular because, according to the activists' press release, it and its subsidiaries commanded more than

15 percent of the student loan industry at the time.[36] Chase was a prime target for the activists because the bank's predecessor was found to have used thirteen thousand slaves as collateral for past loans. Another target was Wachovia, which had disclosed that the banks that preceded it in the nineteenth century owned more than seven hundred slaves; Wachovia soon after announced $10 million in support for minority scholarships and black-owned businesses.[37]

The reparations activists, led by the Restitution Study Group, also started a nationwide boycott of the Aetna health insurance company. During "open enrollment," the window of time in the fall of each year when employees can change their insurance coverage or provider, pro-reparations activists widely circulated an e-mail imploring, "Do Not Use Aetna as Your Health Insurance Carrier. Aetna helped finance slavery in America by writing insurance policies on the lives of enslaved Africans."[38]

Second, Farmer-Paellmann and the seven other plaintiffs sought to persuade the court of their hereditary standing and to counter Judge Norgle's assertion that they "merely alleged" a genealogical relationship to enslaved men and women. In March of 2004, the eight plaintiffs filed a second, narrower reparations case in New York against three companies: FleetBoston, Lloyd's of London, and R. J. Reynolds, claiming that these companies "aid[ed] and abet[ted] the commission of genocide" by providing insurance and financing for slavers. These companies were targeted for their role in insuring and bankrolling the sugar and tobacco trades that were dependent on human bondage. In this case, the plaintiffs sought $1 billion in damages.

The plaintiffs responded to the court's argument that they lacked standing to bring the slavery reparations case with genetic genealogy test results. Farmer-Paellmann turned to the African Ancestry company. Moreover, she and her fellow activists appreciated that genealogy was important evidentiary support for their case. Farmer-Paellmann and Kittles had crossed paths over the last decade in New York City and Washington, DC. Kittles also recalls meeting the activist in Chicago, where there was a large contingent of N'COBRA activists.

In 2003, Farmer-Paellmann visited African Ancestry's offices in the District of Columbia. Paige recalled how the visit took place: "[Farmer-

Paellmann] called and we made arrangements. . . . She came to the office. It was early on [when African Ancestry was getting established]. She was definitely working on reparations. I remember Deadria explaining her work and the relevance of ancestry and the vision that she had for reparations. . . . She bought a test that day. She came with [fellow reparations activist] Annette [Harrell-Miller] and they both bought tests. We definitely weren't in a [financial] position to donate them." Kittles would analyze the samples. "Before I left Howard [for Ohio State University], we did her test. And, she was one of the first who talked about using the test as a prerequisite . . . to prove your authenticity . . . to suggest that you should receive reparations."

While African Ancestry would be an important collaborator, in our conversations both Paige and Kittles expressed reservations about reparations activism. Paige stressed that she "hasn't significantly aligned with any of these efforts." For his part, Kittles admits to some skepticism about this strategy: "Gina and I used to talk. . . . And I used to say, 'Let's not try to get connected too deep with this because I [am] just a scientist.' . . . I don't necessarily believe that we should be asking for anything. . . . That whole issue of going to court and suing banks . . . I would have just taken another approach."

African Ancestry's DNA tests linked the plaintiffs in the slavery reparations suit to ethnic groups and/or nation-states on the continent of Africa. While genetic ancestry testing has been used in family and criminal courts, this would be the first time it would be used in a civil suit to infer African ancestry. Genetic ancestry test results associating Farmer-Paellmann with the Mende people of Sierra Leone, another plaintiff in the group to Niger, and a third to The Gambia, among other contemporary locations on the African continent, were submitted as an evidentiary retort to the standing doctrine. In an interview, Farmer-Paellmann pronounced that she felt a "major void" in her life before she received her genetic ancestry results from African Ancestry.[39] She continued that the financial damages she and the seven other plaintiffs sought were for self and community "repair"; "we want to be whole people," she said.[40]

The DNA data, the plaintiffs argued, confirmed their genealogical connection to Africans ensnared in the slave trade and, therefore, were evidence of their legal legitimacy as aggrieved parties that should be recognized by the court and compensated by the corporate entities that profited from slavery. Indeed, Farmer-Paellmann described the genetic inference provided to her by African Ancestry and submitted to the court as evidence as a "direct link."[41] The plaintiffs' filing declared that "DNA testing has proven beyond a doubt" that there was a fiduciary relationship between the plaintiffs' ancestors and the defendants' companies.[42]

The plaintiffs' submission of DNA evidence as proof of their ancestry accompanied a new legal strategy that took its cue from international human rights law, including prescriptions against genocide. As Farmer-Paellmann put it, "Black people today can only identify with Africa, a continent. That's because our ethnic and national groups were deliberately destroyed to enslave us. We can prove this injury."[43]

In attempting to expand the legal terrain for the case to international law, the plaintiffs' attorneys hoped to detour around the statute of limitations, considered inapplicable to crimes against humanity such as genocide.[44] The argument that transatlantic slavery fell under international law and human rights ethics was bolstered a year before *Farmer-Paellmann v. FleetBoston* was filed, at the 2001 World Conference Against Racism, Racial Discrimination, Xenophobia, and Related Intolerance, in Durban, South Africa. At the conclusion of the conference, more than 150 nations signed a document that asserted the transatlantic slave trade was a crime against humanity. Reparations activists intended to capitalize on this accord. In her writings on the reparations movement, Biondi noted that activists hoped that such a declaration would make the United States "more vulnerable in legal action" to claims based in international law.[45] The difficulty would lie in whether or not this argument could be persuasively pressed for a case about the political economy of chattel slavery that, while deplorable and inhumane, was not illegal when in practice.

Farmer-Paellmann's fervor for genetic genealogy testing led her to establish the Organization of Tribal Unity (OTU), a nonprofit organization that describes its mission as being "to create a network of those who have restored their African ethnic and national identities through DNA testing." In 2006, the OTU initiated an online petition to nominate African Ancestry's Kittles, whose genetic genealogy tests had been submitted in the Farmer-Paellmann reparations suit two years prior, for the Nobel Prize. The petition circulated among reparations activists and black cultural nationalist communities. It read:

> We, the undersigned, propose that Dr. Rick Kittles be nominated for the Nobel Prize for his profound contribution to the field of genetic research. Dr. Kittles, a 40-year-old geneticist descended from enslaved Africans, has earned this honor and recognition for his original DNA research and analysis that is repairing the effects of 450 years of slavery[-]related ethnic cleansing committed against people of African descent. His unique method compares genetic sequences to restore ethnic and national identity—two of the most fundamental human attributes. Prior to Dr. Kittles' groundbreaking work, this information was inaccessible to millions of descendants of enslaved Africans. For using science to "unlock the door of no return," Dr. Kittles deserves the greatest honors and recognition the world has to offer. We propose and support his nomination for the Nobel Prize.

It is worth underscoring a few sentiments in this petition. The "door of no return" refers to the final port or path Africans traveled before making the transatlantic Middle Passage into slavery. If one visits a slave castle today in Africa, such as Elmina on the coast of Ghana, there is a sign leading from the fortress to the shore and marked the "door of no return." The petition drew on this symbolism. Second, Kittles is positioned as revealing information that was not only lost but perhaps also *concealed* from people of African descent. Also, shifting the history of slavery into international law with the post-Durban discourse, the petition positions identity as a human right. In their second attempt at the reparations lawsuit, the plaintiffs stressed that the defendants contributed to the destruction of their national and ethnic identities. Farmer-Paellmann and the other plaintiffs further affirmed that genetic genealogy testing offered a "direct connection" to their ancestors who had been "kidnapped, tortured and shipped in chains to the United States."

Legal scholar Kevin Hopkins, an advocate of slavery reparations, wrote in 2001 about the possibilities of bringing a reparations case to court, highlighting the fraught issue of establishing legally recognized identities. Hopkins predicted that figuring out a way to determine who was eligible for restitution would prove difficult. He weighed the merits of genealogical research, the "one-drop rule," and what he termed "genetic mapping." Hopkins noted that the dissolution of families during slavery and the lack of written records made the reconstruction of a genealogical pedigree very difficult despite new technological developments. A capacious idea of blackness that would include the class of persons with any and all traces of African ancestry was also deemed inadequate because this strategy would not bracket out blacks who are not the descendants of slaves or those blacks who were able to pass for white. Lastly, Hopkins considers what emerging applications of molecular biology would augur for the legal struggle for reparations. He stressed that "DNA testing would provide the most scientifically accurate information for determining the eligibility of Black Americans for slavery reparations" but still "has limitations."[46] These limitations include the fact that genetic evidence alone is inconclusive, in part because the reference databases were incomplete, and these data must therefore be considered alongside other sources of information. Hopkins's analysis would prove to be prescient and, moreover, was cited in this second decision from Judge Norgle.

In March 2005, in a lengthy decision of over one hundred pages, Norgle dismissed the plaintiffs' second case—once again primarily on the basis of standing. The judge maintained that the genetic genealogy tests did not sufficiently establish a relationship between deceased slaves and the signatories to the class-action suit. Norgle noted the strengths and weaknesses of genetic genealogy as hereditary evidence as compared to other forms of proof. Norgle reasoned,

> there may well be no perfect method of determining exactly who is a descendant of a slave, and thus a member of the group entitled to receive reparations. Genealogical research "often fails to provide significant information about a person's ancestry." The blood, or "one-drop," test (whereby anyone with any trace of African ancestry is deemed part of the group entitled to receive reparations) "fails to differentiate between descendants of US slaves and those of other nationalities with African heritage." Genetic mapping, or DNA

testing, is more promising than the above two methods, but "alone is insufficient to provide a decisive link to a homeland."[47]

Invoking both the problem of the incomplete archive and the pernicious logic of hypodescent (i.e., the one-drop rule), Norgle concluded that the only suitable means to establish "a decisive link to a homeland" was DNA evidence that could show an uninterrupted, definitive line of ancestry from an exploited former slave to an aggrieved present-day descendant or descendants, *and simultaneously* a direct line of capital gained from an accused corporation to expropriated laborers and their offspring. Genealogical links were not evidence of personal or familial injury. And genetic genealogy test results, the court surmised, neither supplied the plaintiffs with genetic standing nor substantiated their claims of damage and injury as the descendants of enslaved men and women. Following Norgle's second decision against them, the plaintiffs' sole remaining legal option was to appeal to a higher court.

And appeal they did, to the Seventh Circuit Court of Appeals. Farmer-Paellmann's legal team at this stage included lawyers Carl Mayer, the Yale Law School–educated reparations activist Roger Wareham, and Bruce Afran. The plaintiffs' attorneys requested that the panel reinstate the reparations class-action suit that had been dismissed by the lower district court. Attorneys for the plaintiffs also argued that Norgle dismissed their suit prematurely and failed to fully consider all the evidence before him. The all-white three-member panel of the Seventh Circuit appellate court composed of Judges Richard Posner, Daniel A. Manion, and Frank H. Easterbrook was empaneled to weigh the admissibility of the genetic-ancestry-testing evidence. (Appellate judge Ann Claire Williams, who is African American, recused herself from the case.) To the appeals court fell the task of deciding whether to send the case back to the US District Court for reconsideration or, alternately, to endorse Norgle's 2005 decision to throw the case out altogether.

The writing was on the wall about where this court's sympathies lay on the day the plaintiffs' lawyers petitioned to move forward with their case and offered oral arguments. The plaintiffs sought to have the case kicked back to the lower court so that Judge Norgle would be compelled to more deeply examine the evidentiary details and legal arguments of the plaintiffs' case. During deliberations, a skeptical Posner remarked,

"If you think you've been wronged, it shouldn't take 100 years to investigate the conduct of Aetna, Lehman Brothers, and the like. There are a lot of people living today whose parents were wealthy in the nineteenth century who have nothing."[48] This remark foreshadowed the opinion to come. After deliberations, the three-judge panel decided to throw out the plaintiffs' claims. Without the option of sending the case back to the lower court, this round of political sparring came to a close. The case was struck down on appeal by the three-member appellate court, in a decision penned by Posner, and again on the basis of standing as well as the issue of jurisdiction (i.e., whether this case fall under the purview of the judicial branch or one of the other two branches of government), and the statute of limitations (the anticipated issue the plaintiffs had hoped to get around by invoking international human rights norms).

At a time when antiracist civil rights politics of the type practiced in the late twentieth century may have reached a limit in terms of efficacy, the reparations movement sought and found a new vehicle for trying to face down racial inequality that was deeply rooted in the legacy of slavery—a legacy that has been deeply denied. The turn to reparations through the discourses of human rights and international law posits the reimagining of a persistent demand, newly articulated. The turn to genetic technologies for these goals—the turn to reconciliation projects—changes the terrain of racial justice activism as well. Reparations suggests a shift from "victim" to "creditor." The reparations activists' genetic-ancestry-testing strategy doubles down on this new narrative by suggesting science in the place of sentiment and forensics as a counter to denial.

With reconciliation projects, the insights of genetic science are applied to the discovery or confirmation of ancestry in the hopes of securing forms of social inclusion and measures of social justice, including rights and reparations. But to what extent can DNA identification be efficacious for racial reconciliation? What might be the consequences of the genetic mediation of African diasporic cultural politics that have historically involved social movement tactics and civil rights organizations?

In the case of the reparations class-action suit, there is the problem of hereditary or genetic standing manifested as a gap between how the courts and the plaintiffs respectively interpret relatedness. Tort law requires the succession of capital, matrilineage, and patrilineage to constitute a claim that is "legally cognizable"; that is, it demands an argument that moved through the narrow channels of legal logic rather than traveling through the broader moral currents of the infringement of human rights and the creation of a system—chattel slavery—that constrained a portion of the US population as a caste, with deleterious effects even into the present day.[49] The reparations plaintiffs, on the other hand, introduced genetic genealogy tests into the litigation not only to demonstrate hereditary injury but also to highlight the "social death" inherent in the chattel slavery system.[50] "The injury that we're focusing on," Farmer-Paellmann proclaimed in an interview with the Australian media,

> is the loss of our, the destruction of our ethnic and national groups. African-Americans today do not know who we are. That is a human right, to know who you are. . . . There are now DNA tests available where we can determine the precise ethnic and national groups we come from in Africa, so we're able to trace ourselves back to the slave trade and determine who underwrote those slave trading expeditions, which nations, which companies supplied whatever resources necessary to brutally enslave my ancestors.[51]

Here she points to what political scientist Adolph Reed critically describes as the "symbolic" and "psychological" components of reparations politics, which, he writes, "center on public acknowledgment of the injustices inflicted on black people historically in this country."[52] For Farmer-Paellmann, the loss of identity that accompanied racial slavery is a human rights issue. For Reed, even if it "promotes public education about the real history of the United States," this form of reparations discourse is part and parcel of "the Clintonoid tenor of sappy public apologies and maudlin psychobabble about collective pain and healing."[53] He contrasts these symbolic and psychological components of reparations with material ones that include direct remedies to current-day disenfranchisement and discrimination in housing, education, political participation, and employment.

Farmer-Paellmann and others understand themselves to have taken a different kind of material course. They believe that they have found in genetic ancestry inference a vehicle of racial repair both within and beyond the courtroom. Plaintiffs in the slavery reparations case employed genetic genealogy evidence in order to elicit recognition of their injuries and experiences as a class; they sought to repair a historical injustice by virtue of their shared status as descendants of a slave caste, a collective. The claims of the plaintiffs in the reparations case, while couched in the language of genetics, relies on a sense of connection forged from the principal of "linked fate" (the political scientist Michael Dawson's term) and community-derived forms of kinship. They layer genetics on top of political and historically shaped assumptions about a diasporic black community. Thus, for the plaintiffs, genetic genealogy testing is additional proof of a connection of which they are already certain.

One could conclude that genetic genealogy testing, being of little legal value as a means of securing reparations, is a failed strategy for reconciliation projects. However, the plaintiffs showed considerable ingenuity in putting genetic genealogy testing to unprecedented use and, in so doing, opened new areas in the struggle for racial equality, while also extending a generations-long campaign for reparations that will not abate. The conversation reignited by Ta-Nehisi Coates in the summer of 2014 is an indication that the issue will not go away until some measure of reconciliation is accomplished.

Of course, because genetic evidence is shaped by our aspirations for it and the context in which it is deployed, this strategy comes with the possibility of ironic pitfalls. Writing in the *Black Star News*—a New York City–based independent news service focusing on issues affecting African diaspora communities—Winston Munford speculatively mused that should African Americans be awarded reparations from a national or international entity, another facet of genetic ancestry testing might be deployed. He suggested that the pendulum could swing from discerning differences to highlighting similarities. Munford writes that "people of many races will trumpet any [genetic genealogy] results showing they or their families have a smattering of African blood in them and they will now say that this qualifies them for a share in [r]eparations settlements."[54]

Even though the plaintiffs in the Farmer-Paellmann case ultimately failed in their bid for reparations, they may have gained ground in other ways. As reparations activist Conrad Worrill put it while awaiting the 2004 decision on the case, "History is being made, regardless of what this judge does."[55] Noted sociologist Joyce Ladner agreed that even in failure the case will have "educate[d] the larger public about the role that the institutions played in slavery."[56] Legal scholar James Davey puts a finer point on this perspective, offering that the reparations case may be considered a success in the sense that it heightened "awareness of the issues beyond the confines of the courtroom, in wider social, legal and political fora."[57] To be sure, the debates about reparations reignited by Coates in the pages of the *Atlantic* can be seen as a passing of the baton to a new generation of intellectuals. And more to Davey's point, this airing in the court of popular opinion occurs in the context of the increased ubiquity of DNA, with the legal gate-keepers and the public alike being more exposed to the influence of genetics in numerous aspects of their lives. In this task, Farmer-Paellmann and her allies surely succeeded in keeping the drumbeat for reparations alive.

In a 2010 op-ed in the *New York Times*, Harvard's Henry Louis Gates Jr. implored readers to end "the slavery blame-game" and turned to recent advances in digital history and genetic ancestry testing as offering a way out. However, Gates's DNA project, from *Faces of America* to *African American Lives*—in making slavery vivid and relevant in the present—might be seen to work *within* a reparations framework. Genetic ancestry projects offer a new, if not always successful, strategy for racial reconciliation as part of the sweeping social life of DNA.

Farmer-Paellmann's interest in restitution for the unpaid labor of slaves led her into the world of genealogy. In 2004, a *Chicago Tribune* article linked the then-burgeoning interest in conventional and genetic genealogy to the slavery reparations movement. Indeed the use of genetic genealogy in the reparations movement has been credited with spurring broader interest in DNA testing and even with the revival of interest in African origins. Reporter Dahleen Glanton wrote that "African-Americans are exploring their past in search of answers for the social problems blacks face today." "Just as Alex Haley's book 'Roots' and the TV miniseries based on it ignited an African pride movement," Glanton continued in the *Chicago Tribune*, "the call for reparations . . . has sparked a similar movement 30 years later."[58]

The article goes on to describe the prominence of other roots- and reparations-related activities, including the actions of a middle-aged white couple, Rodrigo and Sheila Fonseca, who participate annually in a cultural event in New Orleans that marks the tragedy of the Middle Passage. Rodrigo identified another process still: "People our age can claim that we were too young and that we didn't do anything. That's true, but our ancestors probably did evil. At some point we have to repent and there should be atonement. For me, this is a reconciliation process for what my ancestors did."[59]

EIGHT

DNA Diasporas

> For those African governments reluctant to provide dual citizenship to Diasporans who cannot legally establish a line of descent, the African Ancestry test offers a means of providing scientific evidence of descent.

Genetic ancestry testing now plays a part in black Americans' desires to solidify ties to the African continent, including a quite literal and practical sense of reconciliation—reconstitution of the far-flung African diaspora. Some people of African descent are not merely seeking a sense of reconnection to an identity or a family lost during the Middle Passage. They further aspire to translate this usable past into bases for present-day affiliation. These new affiliations, which include friendship, DNA kinship, and citizenship, may be further recast into the promise of social, economic, and political collaboration. Like the struggle for slavery reparations, this yearning for affiliation has global implications; the diasporic social network is its modality.

The sociologist Rogers Brubaker has diagnosed a "'diaspora' diaspora" in recent years, arguing that as the term "diaspora" has proliferated, the meaning of the word has been stretched.[1] The hallmarks of diaspora are widely agreed upon, however, and include dispersal of people from long-held geographic homes; the creation of a collective identity or consciousness in response to the experience of dispersal; connection to a place of geographic origin forged through correspondence, tourism, practices of symbolic ethnicity, philanthropy, and political engagement; and the circulation of memories, myths, or imaginaries about the homeland. Diverse diasporas—born of distinct historical, political, and economic push-and-pull factors—share these general contours.

For some, the African diaspora that began in the sixteenth century is "exceptional" among human dispersals of the past and present because it was a forced migration set in motion by the demand for slave labor, and one which spurred the process of ethnogenesis—the substitution of specific African identities for more general collectivities, such as Pan-African, African American, and Afro-Caribbean.[2] As political scientist William Safran maintains, a "*specific* homeland cannot be restored" to the descendants of enslaved men and women.[3] But attempts to restore and return a sense of home persist.

How Africa has been envisioned by its slave descendant diaspora is a topic of debate. At issue is the *ethics* of imagining Africa and diasporic connection: What is the substance of diaspora? Who in the diaspora gets to imagine "home"? How is it imagined and to what ends? Conceptions of Africa that underlie diasporic consciousness may have many sources, including artistic expressions, common experiences of oppression or redemption, communication practices, and today, even DNA.

Notions of diaspora rooted in technologies of kinship such as genetic ancestry testing are "cultures of relatedness" in which biological facts are not the necessary conditions of possibility for social ones.[4] In Carol Stack's classic ethnography, *All Our Kin*, for example, kinship among urban blacks in "The Flats" is based on the exchange of economic resources and caring labor between residents.[5] As Stack shows, members of the community use kinship terms such as "aunt" and "brother," but these categories do not connote blood ties; rather these terms are engaged despite lack of demonstrable genetic links. The forms of sociality fostered by genealogy ancestry testing—both the aspirations for affiliation that inspire its use and the various kinds of relationships it may occasion—are conduits through which diaspora may take shape. The creation of responsibilities and rights, and forms of exchange across the African diaspora, is a common end-result of genetic ancestry practice, with DNA testing facilitating the formation of a diasporic network.[6]

FORGING DIASPORA

I attended a conference of the Afro-American Historical and Genealogical Society that took place at Gallaudet University in Washington, DC. During a "sharing dinner," root-seekers were invited to stand and share highlights of their experiences with the group. Although none at my table availed themselves of this opportunity, we spoke amongst ourselves about our respective genealogical research projects. Seated next to me was Bess (a pseudonym), an African American woman in her fifties who lives near Baltimore, Maryland. I told her about my ethnographic study of conventional and genetic root-seeking—including my preliminary foray into my own family's history. Bess shared that she had been conducting genealogical research on her family for about a decade and had recently received genetic genealogy test results from African Ancestry.

The next morning, I ran into Bess in the hotel lobby, where vendors, including African Ancestry's Gina Paige, were setting up their displays for the day. "I have something for you," Bess told me. As we sat together in the hotel atrium, she volunteered her genetic ancestry test results, arranged neatly in a binder. A letter from African Ancestry indicated that mtDNA analysis had linked Bess to the Kru of Liberia "plus/or Mende-Temne of Sierra Leone." Her result package also contained a Certificate of Ancestry signed by Kittles, a printout of the genetic markers from which Bess's African ethnicity was inferred, a map of the African continent with Liberia foregrounded, and a flyer advertising *Encarta Africana*, a CD-ROM encyclopedia, at a discounted rate.

Bess explained to me that she wants to "do something" with her results, like perhaps "travel to Africa." Curious as to which of the two possible ethnicities suggested by African Ancestry was most compelling to her, I asked Bess whether she planned to visit Liberia, neighboring Sierra Leone, or both countries in the future. "My sister was married to a man from Sierra Leone; his name was Abdul," she replied obliquely, intimating that she would likely travel to the natal home of her deceased brother-in-law. "When will you be ready to travel to Africa?" I asked. "After I get back further in time [with my genealogical research]," she responded. As is common with other root-seekers who make use of genetic ancestry testing, Bess assumed a role in determining her test's significance and its potential import. Yet her intentions for how the test results would be utilized in her life were already apparent. A visit to Sierra Leone was likely in Bess's future, inspired by both a deceased family member and genetic ancestry testing. Her intention to engage in practices motivated by the findings she received from African Ancestry *after* she advanced with her conventional genealogy effort underscores the interpretative work that commences following the receipt of genetic genealogy results. This more deliberative process can involve root-seekers' efforts to align the DNA analysis with other information about their ancestry as well as with their expectations, prior experience, or existing relationships.

Another root-seeker, Marvin (a pseudonym), also used Paige and Kittles's service and found diaspora close to home. The recent family reunion of Marvin, a genealogist from the southern United States, featured an appearance by someone he described as a "genetic kinswoman." Some months prior, Marvin had purchased a genetic genealogy kit from African Ancestry that associated him with the Mbundu people, the second-largest ethnic group in the south-central African country of Angola. Marvin shared his results with a friend, who subsequently put him in touch with Gertrudes, an Angolan immigrant neighbor of Mbundu ethnicity. At their first meeting, Marvin recalled Gertrudes as being "very accepting." He continued, "She said that one of her passions is to connect with African Americans and tell them about their history in Africa and to let them know that, as she always says, 'we are one.' [She believes that] there is a disconnect between African Americans and Africans, and she's trying to bridge the gap. One of her missions is to connect with more African Americans [and] teach them about Africa." They took different approaches, but both Marvin and Gertrudes sought, in their own way, to reconcile the African diaspora.

Gertrudes subsequently invited Marvin to attend a celebration of the thirtieth anniversary of Angola's independence from Portugal, hosted by the voluntary association for immigrants from the African country that she helms. Here, Marvin and a cousin who attended the party with him felt accepted by the larger Angolan expatriate community as well. "Once we told everyone there that our family came from Angola, they all said, 'Welcome home. You're home now.' They even made me and my cousin get up on the dance floor. You know, they do a ring dance? . . . They told us, 'You gotta come dance. Dance for your homeland!'"

In turn, Gertrudes would attend Marvin's family reunion some months later. "Her presence was powerful," Marvin recollected. "[She talked] about the importance of us coming together as a group of

Africans. She expressed that we are all Africans and that Europeans try to divide us but now we must come together. And she also told our family some very interesting facts about the Mbundu people. And that was awesome, just for the family to hear about the people we descend from . . . directly from an Mbundu person. It was very powerful. She had the full attention of the whole family. Everybody was just sitting there in awe of her presence. . . . It was uplifting and powerful just to hear her tell us something about our African roots."

DIASPORIC RESOURCES

The social exchange carried out between Marvin and Gertrudes points to how genetic ancestry testing circulates as a "diasporic resource."[7] As anthropologist Jacqueline Nassy Brown describes, diasporic resources can include "cultural productions such as music, but also people and places . . . [and the] iconography, ideas, and ideologies" of one black community that are employed by another as formative schema for political consciousness, collective empowerment, and identity formation.[8] In Brown's work, the concept describes, for example, how knowledge of a historic era such as the civil rights movement of the mid-twentieth-century United States circulated globally via the media, popular culture, and social networks to become an important touchstone of self-determination for blacks in Liverpool, England, in the 1990s.

In the context of ancestry testing, the concept of diasporic resources elucidates how genetic information occasions the weaving of a social mesh between African communities and their dispersed members, even in the absence of specific kinship ties. An imprecise pedigree connects Marvin and Gertrudes as "genetic kin," as "DNA Mbundians," "DNA Angolans," and "Africans."

There are also economic and cultural resources at stake in the DNA diaspora. The international press has reported on attempts by leaders on the continent to have prominent African Americans "correctly" affiliate with their DNA kin on the continent. After the actor Whoopi Goldberg was inferred to be related to Papel and Bayote communities of Guinea-Bissau using African Ancestry testing on the *African American Lives* television show, the Associate Press reported that "the government of one of the world's poorest nations" wanted her to come "home." "She's our daughter. She's ours," a government official said.[9] While a specific request for support from Goldberg was not conveyed in the AP story, this implication was clear. For to say "she's ours" is to try to enlist her in diasporic obligations.

It was also on *African American Lives* that we learned that media mogul Oprah Winfrey discovered that she has been affiliated with the Kpelle people in the country of Liberia via Kittles's genetic ancestry test. The rub was that two years earlier, Winfrey had proclaimed publicly that genetic genealogy testing showed her to be Zulu. "I always wondered what it would be like if it turned out I am a South African. . . . Do you know that I actually am one? I went in search of my roots and had my DNA tested, and I am a Zulu," she declared. Having established her Leadership Academy for Girls in South Africa in 2007, the appeal that a genetic association with the Zulu-speaking Nguni people held for Winfrey was clear. After she was subsequently affiliated with Liberia in a televised "reveal," Winfrey conveyed that she still feels "at home" in South Africa and wished that she "had been born [there]."[10] However, soon after, Winfrey backed an effort to "place Liberia on the front burner of donor and philanthropist attention."[11]

Isaiah Washington became one of the few US citizens to ever be granted dual citizenship in Sierra Leone. This formal affiliation with his "homeland" was allowed based on his mtDNA genetic ancestry test—and undoubtedly also his prominence. As is always the case with genetic genealogy, Washington's status as a "DNA Sierra Leonean" was owed to the presence of genetic samples from people of that country in African Ancestry's reference database. Whether this is a few people or hundreds of people, we cannot know for sure. What we do know is that the authenticating DNA markers residing in the proprietary African Lineage Database, and in those of other DTC genetic testing companies, are their own kind of diasporic resource. This fact in and of itself may create a sense of obligation to the ethnic groups with which one is associated. Although Washington never mentioned this specific debt to Sierra Leone or the African continent, he is clear that much of his life's work lies there with the creation of social welfare programs, the recent fight against the Ebola epidemic, and other endeavors.

While diasporic resources are transnational, they may be unevenly distributed. Unlike the circuits described by Brown in which music and affects like "freedom" and "moral authority" are the stuff of the black diaspora potlatch, the celebrity nature of ancestry "reveals"—which, as Paige explained, derive from the early days of the company—highlight yawning inequality across the network and foreground economic exchange explicitly.

BUILD, BROTHER, BUILD . . .

The DNA diaspora is also being forged through organizational efforts. The continuing work of the Leon H. Sullivan Foundation (LSF) exemplifies this. The LSF is a nonprofit organization with roots in the civil rights movement. It is named for an influential African American Baptist preacher and activist who, for decades before his death in April 2001, was a longtime crusader for racial equality and economic advancement for blacks. Sullivan was also a zealous promoter of African diasporic cooperation and mutual aid. He would liken himself to Robin Hood: "I make no excuses about it . . . I take from the rich and give to the poor. It is high time at last for the world to wake up and do more for Africa."[12] In recognition of these efforts, during his lifetime he was granted (mostly titular) citizenship in Cote d'Ivoire, Gabon, and The Gambia, in honor of his social and economic development work in each of these countries.

Beginning in 1991, the LSF inaugurated the Leon Sullivan Summit, a gathering of elites from across the United States and the African continent. The conferences consist of dialogues between politicians, bureaucrats, and business leaders from varied African countries and their American counterparts. Prominent American attendees at past summits have included George W. Bush, Bill Clinton (an honorary cochair of the board of the LSF), Condoleezza Rice, Colin Powell, the Reverend Jesse Jackson, and business leaders such as Paul Wolfowitz, a former World Bank head.

"Inspired by Rev. Leon H. Sullivan's belief that the development of Africa is a matter of global partnerships," the first summit took place in Abidjan, Côte d'Ivoire, in 1991.[13] Subsequent summit gatherings took place in Libreville, Gabon (1993), Dakar, Senegal (1995), Harare, Zimbabwe (1997), Accra, Ghana (1999), and Abuja, Nigeria (2003, 2006). The 2008 meeting, held in Tanzania, would mark a turning point for the organization, culminating in an initiative that reflected Sullivan's spiritual calling as a bridge builder.

Leon Sullivan was born in Charleston, West Virginia, in 1922. As a child, he was deeply influenced by his religiously devout grandmother, Carrie, who raised him for several years, while his mother worked in Washington, DC. Maturing quickly under the tutelage of Reverend Moses Newson of the First Church of Charleston, Sullivan became a preacher at the age of seventeen. By eighteen, he was pastor of two churches. He served in the ministry in West Virginia for a total of four years before moving to New York City to attend Union Theological Seminary; Sullivan joined the seminary at the urging of the Reverend Adam Clayton Powell Jr., a fellow preacher, who served as a US congressman representing New York City's Harlem community for more than two decades. The activists and intellectuals of Harlem caught Sullivan's attention soon after his arrival in 1944, showing him examples of how to combine his religious zeal with his passion for social justice. While in his early twenties, Sullivan became involved in some historic events and important institutions. He was a leader of the March on Washington movement, and an assistant pastor at the historic Abyssinian Baptist Church.

In 1945 Sullivan took a post as deacon at the First Baptist Church of South Orange, New Jersey, where he began to experiment with using the church to develop and foster community-service programs. As he wrote in his autobiography, *Build, Brother, Build*, Sullivan dedicated himself to "the problems of the community and did all [he] could to assist colored youth there with employment opportunities, as well as scholarships and other educational assistance programs."[14] In 1950 he became pastor of the Zion Baptist Church in Philadelphia. With Sullivan at the helm, Zion became an "urban Christian center," whose mission included ministry as well as programs such as a day care center, adult education courses, and a credit union.[15] Because the preacher believed that economics were key for black uplift, employment and job training remained a priority. He and other ministers organized a series of successful boycotts in Philadelphia in the 1950s and early 1960s as leverage to open up job opportunities and eliminate employment barriers for blacks in the city. In 1963 he launched his Opportunities

Industrialization Centers (OIC) to provide job-skills training. By the late 1960s, there were more than seventy-five centers across the United States; the program even went international, with OICs established in eighteen countries, including several in African states.[16]

Widely hailed for his ingenuity in creating economic opportunities, Sullivan would become the first African American member of the board of directors of a major multinational company when he joined General Motors. On the GM board, he led an effort to get the company—which has business interests in South Africa and was the largest employer of blacks there—to divest from the country until the apartheid regime had ended.[17] In recent history, he is perhaps best known for developing "The Sullivan Principles" in 1977, an explicit anti-apartheid effort that established ethical, egalitarian codes for how non-white workers should be treated by US-based multinationals conducting business in South Africa.[18] Sullivan's corporate activism was inspired by his experience traveling in the country. He recalled the experience: "In 1974 I met with hundreds in South Africa and I realized that apartheid was sinful. . . . When I was getting on the plane to go home, the police took me to a room and told me to remove my clothes. A man with the biggest .45 I'd ever seen said, 'We do to you what we have to.' I stood there in my underwear, thinking, 'I'm the head of the largest black church in Philadelphia and I'm on the board of directors of General Motors. When I get home, I'll do to you what I have to.' "[19]

The self-proclaimed African American Robin Hood drew attention to South African social and political conditions. As leverage, the Sullivan Principles held nothing but "moral capital" (to borrow historian Christopher Leslie Brown's phrasing); yet by 1986, 183 firms were on board and the principles had been expanded to "include direct challenges to the law and institutions by which apartheid [was] maintained."[20] As the *New York Times* put it, "Virtually all American companies doing business in South Africa . . . abide[d] by [Sullivan's] guidelines."[21] In 1991 George H. W. Bush honored Sullivan with the Presidential Medal of Freedom.

In 1998 Sullivan recalled how his vision for transnational African diasporic meetings evolved in this way: "I had heard and heeded a call from God and from Africa and from African Americans and others of the black diaspora to try at least to unite people of African heritage with Africa, *to make a link, to build a bridge*. . . . The building of this bridge would lead to a series of landmark African-American summits."[22] It was at the eighth such summit, in Arusha, Tanzania, in June 2008, that the diasporic connection which Sullivan had fostered for decades was transformed from an imagined transnational community based upon shared historical experience into a bridge soldered with biological ties. At the close of this summit, the participants, including the African American actor and genetic genealogist Isaiah Washington, approved a resolution to formally recommend that US blacks purchase DNA testing from African Ancestry. These root-seekers would then be encouraged, but not required, to embark upon philanthropy and economic development projects in the countries to which they were genetically matched.

A BRIDGE TO AFRICA

The Arusha declaration was an outgrowth of work the Sullivan Foundation was already carrying out. In 2004 it began to focus on "the promotion of dual citizenship" as one of the cornerstones of its work.[23] Gregory Simpkins came to this work through a meeting with Hope Sullivan Masters, the daughter of the late Leon Sullivan and LSF president, at an organizing meeting for the African American Unity Caucus. "Later, they were looking for someone to do some work . . . and they asked me to come work with them. I had found out that Reverend Sullivan had been interested in dual citizenship, had been given dual citizenship in Cote d'Ivoire and Gabon and a couple of other countries as a result of the summits, and I thought, 'Let me look into that.'"[24] After Simpkins joined the LSF as vice president, a presentation by a State Department official at a subsequent Leon Sullivan forum in Washington, DC, confirmed to him and his colleagues that US citizens were permitted to hold dual citizenship as long as they did not renounce their American nationality, or become embroiled in a conflict against the United States. "Looking at . . . what [Sullivan] had done with the dual citizenship," Simpkins, along with Anthony Archer, a California-based attorney, and others in the LSF, wanted to use genetic genealogy testing to achieve the same end. "What we wanted to do was to engage people in the diaspora and I think to be able to do that you have to have some ownership, some connection . . . the DNA test certainly provides that now . . . [we] felt that blood tie is a connection you have to have. [But] some countries . . . Nigeria, Gabon, they don't require that at all. A country like Sierra Leone [where] Isaiah Washington . . . [and] Andrew Young [have genetic ancestry matches] . . . they do prefer that. I don't know that they actually require that, but they do prefer that."[25]

Simpkins had first met African Ancestry's Kittles in 2003. Hearing of Kittles's work, he eagerly made the quick trip across Washington to visit the geneticist and assistant professor in his lab at Howard University. This encounter would stay with him. In 2007, Simpkins proposed genetic genealogy testing as a "mutually beneficial" partnership between the Leon Sullivan Foundation and African Ancestry. He described the relationship further: "It was a collaboration, because the Foundation was in a sense recruiting people, giving them a rationale for taking the test. The test built interest in the summits and other programs of the foundation. So, it was a symbiotic relationship."[26]

For the 2008 meeting in Arusha, at which genetic genealogy testing would be recommended, Archer presented a report, drafted with the assistance of African Ancestry's Gina Paige, describing the goal of the company's DNA testing partnership with the Sullivan Foundation:

> Recently, the African Ancestry company has developed a DNA test that can provide genetic linkage to African ethnic groups and the country in which that ethnic group now resides. As the company states, the test indicates where the particular strain of an ethnic group is located today and not where they were when ancestors were taken off the continent. This test does not establish an identifiable link to a specific African country or countries, since most tests reveal multiple ethnic heritages located in more than one country. There are 27 countries providing ethnic matches to those taking the African Ancestry test, but the leading countries with links identified by the test are: Nigeria, Cameroon, Guinea Bissau, Liberia, Ghana and Sierra Leone. For those African governments reluctant to provide dual citizenship to Diasporans who cannot legally establish a line of descent, the African Ancestry test offers a means of providing scientific evidence of descent.[27]

Paige took meetings with the LSF's leadership to develop this partnership and proposal. "They called me; I went to the office," she remembers. "That was when the idea of incorporating [African Ancestry's genetic genealogy testing] into the summit arose. The idea was to encourage people to purchase it with their registration and then we could reveal everybody" at the Tanzania summit meeting. "A link where people could buy the test kit" was included on the website for the summit. "There was a grand vision," she continued, "but it wasn't part of their plan from the beginning."[28]

As part of this vision, Archer proposed three levels of possible citizenship for members of the African diaspora: symbolic, partial, and active. Simpkins, who is now a staffer on the House Subcommittee on Africa, Global Health and Human Rights, explained the three levels to me in this way: "There are those who just want to be able to go back and forth without a visa, there are those who want to be able to own property and have some of the economic rights of a citizen, but they're not really going to live there, at least not permanently. And then there are those who are going to want to live part of the year or have that option."[29] On the face of the matter, the benefits of this three-tier citizenship scheme accrued most to the relatively wealthy blacks in the diaspora. But there were also real economic benefits that accrued to African nations as well through tourism, real estate investment, tax revenue, and the like.

Simpkins explained his reservations with respect to the varied dual-citizenship plans, and spoke passionately about the limits of diaspora, even when the affiliation is rendered through DNA:

> I don't think it's necessary for all of us to become involved in [African] politics, because if you don't live there I don't think you know what's going on in their politics. The other complication is that in Africa ethnicity is a serious issue. Now, I took the [African Ancestry] test and was half Zeframani and Tikar from Cameroon. I know what that means historically, but I don't have the same feelings as a Tikar about his or her place in society. . . . We're cousins, we're not brothers and sisters. . . . Even when you've got proof, it would take you some time to adjust to the fact that this is your relative now; you're cousins. . . . The feeling of kinship doesn't nor-

> mally follow. It takes time to establish. . . . You can establish kinship with *anyone*. . . . But it's going to take some time; it's going to take some work.[30]

The various posts Simpkins has held in the federal government, in the media, and at NGOs have allowed him to travel extensively in Africa, "from Cape to Cairo," as he put it. Even as Simpkins championed this plan, he was aware of its cultural and technical constraints.

> I think I've been to like twenty-eight countries. . . . And I don't try to pretend that I'm African. I'm a descendant, but I'm not an African, I am an American. You know, I get along much better, the people who come and pretend that they're exactly the same; they have a problem, because you're not.[31]

Yet most "diasporic Africans" likely did not share the sentiments of Simpkins, who had spent considerable amounts of time on the continent of Africa and was especially well-versed in its history and politics.

From the other side of the transatlantic equation, the LSF needed to quell African collaborators' concerns that African Americans would not feel entitled to become full members of the societies to which they were matched. As the LSF report stated, "the African Ancestry test offers a means of providing scientific evidence of descent." Yet some "African governments [are] reluctant to provide citizenship to historic African Diasporans." The LSF formed an advisory group to investigate the matter. Because, for some governments, African Ancestry's tests did not definitively establish relatedness such that root-seekers should be unequivocally granted rights and benefits in their countries, this committee also considered whether and how its genetic genealogy services could be a technically valid method for extending a "right of return" in African countries for blacks in the diaspora.

Paige pointedly conveyed to me that the "grand vision" never quite made it off the ground "because it wasn't really promoted." She attended the 2008 Sullivan Summit in Tanzania, but there were no attendees to be revealed. "It never came to fruition. I ended up going on that summit. . . . There was no African Ancestry anything. It's a grand vision that I think would have been fairly easy to execute but I am not sure that there was any one person who was responsible for making sure that it happened." Simpkins left the Sullivan Foundation in 2011. He says that the African diasporic DNA initiative "left when [he] left" the organization. However, as I detail in the next chapter, the collaboration did succeed on one weighty occasion.

In the meantime, the DNA diaspora networking continues. African Ancestry has recommend heritage tourism trips to African countries based upon customers' genetic test results. Soon after I received my own genetic affiliation to Cameroon in September 2010, I also received an invitation from the company to participate in a group visit to that country. I did not join the tour. But in December 2010, Paige led a tour of over fifty African Ancestry customers to Cameroon for a ten-day "Ancestry Reconnection Program" designed especially for "African Americans who have traced their DNA to Cameroon."

As Paige recalls, the program got started when Denise Rolark Barnes, publisher of the *Washington Informer*, had her ancestry traced to Cameroon by African Ancestry. This news reached members of a Washington, DC–based, Cameroonian cultural organization called ARK Jammers (ARK stands for Acts of Random Kindness). ARK Jammers proposed an expenses-paid trip to Cameroon for a group of DNA Cameroonians. Travelers only had to cover their airfare. "I started talking to Avline at ARK Jammers. They wanted to take people back to Africa, to their home countries . . . they wanted to start with Cameroon because that's where they are from," Paige said. Although Paige and Kittles have no plans to extend their work into the tourism industry, she "felt compelled to go because I felt that [African Ancestry clients] were going based on my recommendation. I felt that I needed to be there." The first trip of fifty-four people included Howard University psychology and engineering students who "took the tests in advance" and went on the trip "as part of their international exchange" program.[32]

"It was definitely a transformative experience. . . . We met with the prime minister and people in every level of government except for the president. . . . The tour organizers made sure to give everyone an experience tailored to their ethnic group [i.e., Tikar, Bamileke, etc.].

Fifty-four people convened at Dulles [airport] and they were African Americans who traced their ancestry to Cameroon. When we landed in Douala, they were Cameroonian American. After ten days, when we got back on the plane [to return to the United States], they were Camericans. They completely redefined their identities. . . . It was definitely impactful."[33] Although citizenship was not an express possibility, during this tour, root-seekers were permitted to make "formal requests" for Cameroonian "national identity cards." When a second Ancestry Reconnection Program trip was held the following year, ninety-two people attended, including both Paige and Kittles.

Following the Leon Sullivan Foundation's resolve to boost the benefits of genetic ancestry testing, the biennial Sullivan Summits stalled. But in 2010, the organization held a Global African Reunion in Atlanta, Georgia. Organized in late September, just after the close of the UN General Assembly, during a window in which some African politicians would be in the United States, this gathering, described in the next chapter, brought together many of the same kind of stakeholders who had attended past summits: ambassadors, presidents, and other politicians, investors, and other elites from both sides of the Black Atlantic.

Hope Masters, Sullivan's daughter, organized the LSF's programming with a much more lavish touch after her father passed away. This strained the organization's resources. She planned the 2012 Summit for Equatorial Guinea with significant financial support from the country's leader, Teodoro Obiang Nguema Mbasogo, who is a widely criticized authoritarian and who had reigned as president since 1979. But these best laid plans and the organization unraveled in controversy.

After news spread that the LSF had hosted an event in honor of the dictator, support for the organization and the next summit dwindled, with 75 percent of the attendees canceling their participation, the withdrawal of support from corporate backers such as Coca-Cola, and the loss of symbolic support from former president Clinton, whose name and role as honorary chairperson was wiped from the foundation's website.[34] This attempt at diasporic politics had failed, and perhaps too this particular legacy of bridging blacks across the world to which the elder Sullivan had devoted his life. Ghanaian activist George Ayittey explained to the *Washington Post* that there was something deeply "American" about how Masters had endeavored to extend the legacy of her father's mission, noting a critical miscalculation: "For African Americans, the solution for advancing civil rights has come from working within government," Ayittey argued. But he and other Africans "see our government as the problem."[35] As Simpkins advised, one "can establish kinship with anyone," but doing so requires an investment in the time-intensive work of mutual understanding. DNA alone may not be enough.

NINE

Racial Politics After the Genome

> The legacy of slavery, Jim Crow, discrimination in almost every institution of our lives—you know, that casts a long shadow. And that's still part of our DNA that's passed on.
>
> —*President Barack Obama*

In my discussions with genealogists, academics, and others about my research on genetic ancestry testing and the uses to which it has been put, I would inevitably be asked about my own experience. I would always answer honestly, stating that although I found genetic ancestry testing fascinating—which is why I had spent more than a decade studying it—engaging in the practice was not a priority for me in the way it was for many of the people I encountered. As a matter of fact, because of my close association with the traditional genealogists at the Jean Sampson Scott Greater New York Chapter of AAHGS in Harlem, who do archival work and make use of the growing availability of databases of digitized records, conventional family-history research had tended to hold more appeal for me.

As an ethnographer, I was also aware that the frequency with which I was asked, "Have you taken the test?" reflected impressions about my credibility as a researcher. The subtext of the question was: "How can I trust your analysis if you haven't had the experience?" Scholars and reporters often write about social phenomena they have not experienced firsthand. Yet my growing understanding that genetic ancestry testing was not just personal but also societal and political—and my methodology of participant-observation—ultimately led me to embark upon my own genetic genealogy journey.

I had planned to attend the 2010 Leon H. Sullivan Summit, as this would have been the first summit since the passage of a resolution encouraging blacks in the United States to use African Ancestry's testing to form diaspora networks. I hoped to take stock of what this declaration was beginning to yield. Yet owing to fiscal troubles, the 2010 summit did not take place. And the fate of the Leon Sullivan Foundation was becoming increasingly uncertain.

However, the organization announced a Global African Reunion in place of the biennial summit for four days in September 2010. Among the conference registration options was a "DNA Test" plus attendance package. Here was my chance to engage personally in genetic genealogy and to experience the public reveal of my results. In July I swabbed my cheek and sent my genetic data to Sorenson Genomics, a leading DNA testing laboratory based in Salt Lake City, which processes samples for African Ancestry as well as Ancestry.com. In two months, I would receive the results of my MatriClan test.

In August, I received an intriguing e-mail announcement about the Global African Reunion, with the subject heading "A Royal Reveal," from the Sullivan Foundation:

> Our Reveal Ceremony has just become *Royal*
>
> Please join the Leon H. Sullivan Foundation
> and Martin Luther King III
>
> As he learns his ancestral lineage, and that of his father,
>
> The Reverend Martin Luther King, Jr.
>
> . . . FOR THE VERY FIRST TIME THE AFRICAN COUNTRY WHICH CAN CLAIM **THE GENETIC BLOODLINE** OF REVEREND MARTIN LUTHER KING, JR. WILL BE REVEALED.

DREAM NO MORE

As was evident at the 2011 National Urban League meeting that would kick off with an announcement of a 23andMe research project for African Americans—Roots into the Future—in which genetic ancestry testing would be provided for free in exchange for giving a DNA sample to be used in medical research, genetic analysis is being interwoven with the legacy of the twentieth-century civil rights movement. This unlikely conjunction was on full display at the Global African Reunion a year earlier. This was not the "royalty" that many genealogists hope to find in their family tree—some distant king or queen from a precolonial African empire like Benin or from early modern Europe like the House of Habsburg. This was African American royalty. This was royalty from the land of the long black freedom struggle. And I was to receive my genetic ancestry reveal on the same evening.

The Leon Sullivan Foundation's Global African Reunion was held in Atlanta in September 2010. It was here, in a hotel ballroom, that I would learn in a public "reveal ceremony," standing alongside two other African American women—one an employee of the Compound Foundation, launched by singer-songwriter Ne-Yo, and one an LSF affiliate—my genetic ancestry. Kittles presented my results to me along with his business partner Paige, with Isaiah Washington, the evening's MC, beside him. Our three reveals were the opening act. The reveals of three prominent men would be the headlining act of the evening. There would be only kings at the royal reveal.

Paige announced the DNA results to the two women on stage with me. They were visibly moved by the news that "at long last," as one of them said, concluded this stage of their pursuit of African ancestry. My reveal presentation was the last one of the opening act. Paige, who like Kittles was well aware of my ongoing study of their company, turned the microphone over to her partner, perhaps because in my research I had worked more closely with him. Waiting to hear my results, I was worried that my skepticism about genetic ancestry testing—its technical limits and its symbolic excess—would be apparent to the audience and would ruin the experience of genetic revelation and African reconciliation that they had come to enjoy. I had already witnessed the exhilaration of the two women before me and the emotional call-and-response they carried out with the enrapt and excited audience. Was I a fraud? Should a researcher who has scrutinized the ritualized nature of the reveal since the advent of genetic genealogy participate in this ceremony? Because I was well aware that the reveal was among other things a dialogue with an audience, I wondered if I could deliver the emotion that this revelation was expected to elicit.

Because I had developed a professional relationship with Kittles as a fellow academic, the "handover" from Paige to Kittles quelled some of my anxiety. I knew he was aware that I studied the gray area, the nuance and complexity of genetic ancestry testing, and that my reaction would likely reflect that. "Alondra, you are related to the Bamileke people of Cameroon!" I looked at Kittles, smiled widely, and said "Thank you," feeling a bit disoriented. I looked out into the crowd, seeking a theatrical cue. What else was expected of me? The response came when the room swelled with audience applause. My reveal had succeeded. Kittles handed me an envelope containing my Certificate of Ancestry and other details about Cameroon. As I descended the stage, I was patted on the back, squeezed on the shoulder, and hugged by many. The actor Rockmond Dunbar, dressed in an African tunic, offered words of congratulations and encouragement. It was a surreal experience. I experienced in the emotion of these encounters what I had been trying to convey conceptually. It was like the descendants of Venture Smith explained: genetic ancestry analysis provides results to an individual, but it is about so much and so many more.

The headliners—the reveals of not only a son of Martin Luther King Jr. but also the son of Marcus Garvey as well as Carlton Brown, the president of Clark Atlanta University—took place before the same expectant crowd that included Dunbar, as well as former Atlanta mayor and United Nations ambassador and civil rights icon Andrew Young, former president of Nigeria Olusegun Obasanjo, representatives from the administration of Nigerian president Goodluck Jonathan, bureaucrats from across the continent of Africa who had just completed their work at the UN General Assembly, numerous African expatriates, and many African Americans. Although King's legacy has been riven in internecine struggles amongst his children, King III was accompanied on this evening by his sister Bernice. Julius Garvey, a septuagenarian retired cardiac surgeon from Long Island, New York, wore an unassuming gray suit, his father's politics evidenced in a small lapel pin with equal-sized bands of red, black, and green that, to the uninitiated, was as subtle as a Rothko painting. But all in this room knew it to be a symbol of Pan-African politics.

MC Isaiah Washington signaled for Kittles to announce the results. Kittles informed King that his Y-chromosome DNA analysis traced to

Ireland and his mtDNA analysis associated him with the Mende people of Sierra Leone. The same tests inferred that Garvey was connected to Portugal and Spain on the patrilineal side and to Guinea-Bissau, Sierra Leone, and Senegal on the matrilineal side. (I would learn later that King and Garvey had received their respective results long before this evening.)[1] The reveal was for the audience, it was to complete the narrative arc of reconciliation, the ritual.

The headliners were asked to speak to the audience. Standing at the podium with Kittles, Paige, and Washington, and with Hope Masters, the daughter of the late Leon H. Sullivan, standing just behind him, Garvey focused on the patrilineage connection to Europe. Stepping up to the microphone, he reminded the audience of the brutal history of slavery that yielded this genetic result. King, for his part, standing at the podium with his sister, underscored the global family aspect of his particular results. He said:

> For all the years that I traveled to the African continent, I always wanted to know now where exactly do I hail from. So now, we know. The question is what's next. Dr. Garvey shared information that put things in perspective from the [patrilineal] side of our fathers. But it is all connected because our world is a very small place and getting smaller. And somehow we as people of color, not that anyone else doesn't have a special role, [have a special role to play]. But I used to hear our father say that the historian Arnold Toynbee used to say that if there was to be any level of morality brought to the shores of our world, it may have to come from people of color. And so we stand on the brink of history, on the shoulders of so many tonight. And we must quickly, Bernice, my wife, Andrea, and other family members that are here tonight: we must quickly get to the shores of Sierra Leone, quickly connect with the Mende people. And then we ought to do a little more research on the Portuguese side. But from the bottom of my heart again I just say 'thank you' for this incredible connection. Bless you.

These responses highlighted the narrative and contextual framing that is so crucial to the social life of DNA. Genetic markers in and of themselves have no meaning or value; they are like letters on the page of a book for someone who has not been taught to read. But we learn to read the significance of DNA in science labs and at genealogical gatherings. Garvey and King similarly received ancestral inferences to both the continent of Europe and the continent of Africa. Their interpretations gave different tonal emphases, but neither timbre was wrong.

As the scions of historic "race leaders," the participation of King III and Garvey made patent the social justice and social repair motivation of reconciliation projects. Yet, on this evening, there were no exultations of "Up, up, you mighty race!" as Garvey famously encouraged his followers. There was no overt expression of a renewed vision of equality as articulated by Martin Luther King Jr. in his "I Have a Dream" speech. The DNA test revelations played out as the twenty-first-century legacy of the civil rights tradition, and were implied to carry forward this work of transformation and imagination. These reveals nevertheless lacked Garvey's organizational structure and King's prophetic egalitarianism. Indeed, the prevailing social vision on offer this night—through a large projection of the Reverend Leon Sullivan brandishing an African passport that loomed over the ballroom's stage and Washington's wielding that night of his own recently acquired Sierra Leonean passport—was one of escape.

The Global African Reunion encapsulated the vexed nature of reconciliation projects, which are simultaneously atavistic and futuristic. On the one hand, this event conveyed that the racial reconciliation, recompense for past injustice, and diasporic reunion—the resolution of past yet persistent matters—had been reimagined as a technical issue of the highest order. These were cutting-edge solutions to age-old injustices and yearnings; new tactics for new times.

It was vividly evident as well that most sitting with me in this Atlanta ballroom shared the conviction that we were living—and performing—an urgent, different kind of racial politics. In this new time, genetic science can be friend or foe. It can exonerate or convict. It can be a bridge to Africa or underscore the bittersweet European ancestry of many blacks in the United States. What ultimately did I as an individual, and this transnational audience, gain from these six public roots revelations?

As I was poignantly congratulated by countless strangers that evening for having recovered my African ethnic identity, but felt no differently, I was keenly aware of how deeply symbolic this experience was. At the same time, I was very conscious of the fact that DNA holds not only the molecular building blocks of life, but also some of our highest aspirations, for ourselves, our families, and our social communities.

With reconciliation projects, the insights of genetic science are applied to the discovery or confirmation of ancestry in the hopes of securing social inclusion, including rights and reparation. But to what extent can DNA identification be efficacious for African diasporic and/or racial reconciliation? What might be the consequences of the genetic mediation of African diasporic cultural politics that have historically involved social movement tactics and civil rights organizations?

In this book, I have explored the interconnection of genetics, racial politics, and aspirations for social repair at several interconnected sites. Knowledge derived from genetic science has increasingly been used to explain ever-growing aspects of the social world, as demonstrated by the proliferation of genetic genealogy testing. In the face of our growing faith in DNA, caution is warranted. The reconciliation projects described here suggest that it may not be possible to settle political controversies and correct historical misdeeds on strictly technical grounds. Genetic genealogy tests may lack strong enough precision to be efficacious in the courts, or to rule out the possibility of other ethnic or racial affiliations with certainty.

A troubling reality of these reconciliation projects is the fact that the purposes to which DNA is put may be, in the words of Alvin Weinberg, "trans-scientific."[2] A nuclear physicist, Weinberg deployed the idea of trans-science to forestall criticism of potentially dangerous research and to draw a line between politics and "pure science," as he put it. Contra Weinberg, I take it as a given that science and its applications are inherently social (and thus also political) phenomena. But I nevertheless find Weinberg's insight to be of use at a time when genetic science is being asked to solve and resolve myriad issues.

For Weinberg, some questions posed to science—typically metaphysical or moral ones—cannot be answered or resolved by science itself. The inadmissibility of genetic genealogy as proof of ancestry in civil suits—despite its use in other courts and for other operations of state power, such as the reunification of immigrant families in immigration policy, and the expansion of "familial searching" in the criminal justice system, for example—suggests that reconciliation projects may be trans-scientific. In other words, while the practice of scientific inquiry emerges from social and political conditions, it may nevertheless prove incapable of grappling with issues that are essentially ethical, metaphysical, or moral in nature. Some of the questions posed to genetic science may be fundamentally irresolvable through DNA analysis, such as centuries-old and deeply entrenched disputes and debates about racial slavery in the United States. Clearly, the issues, controversies, and questions we pose to science about race and the unsettled past can never find resolution in the science itself.

Those instances in which genetic science fails to fully resolve these concerns suggests that what is sought are not genetic facts as proof of injury or vectors of repair, but rather reconciliation in its fullest sense. The repair that is sought cannot necessarily be found in genetic science solely. DNA can offer an avenue toward recognition, but cannot stand in for reconciliation: voice, acknowledgment, mourning, forgiveness, and healing. These reconciliation efforts also raise interesting and fraught contradictions: they threaten to reify race in the pursuit of repair for injury; they suggest how the pursuit of justice can be easily intertwined with commercial enterprises; they may substitute genetic data for the just outcomes that are sought; and, indeed, they demonstrate well that facts may not, in and of themselves, secure justice.

We also ask DNA to embody some of our loftiest goals for social betterment. Tied to our genealogical aspirations are ambitions for ourselves, our communities, and our world. The initiatives described in this book are taking place at a time when other avenues for social justice and racial repair, such as affirmative action, have been curtailed or lost popular support. Reconciliation projects are one form of response to both the retrenchment following major civil rights strides of the 1950s and 1960s, and also the narrowness of our electoral politics. At the risk of mixing metaphors, in reconciliation projects the double helix works as a spyglass that telescopes back in time, allowing us to see the healing that remains to be achieved in American society. While attention on genetics is today understandably focused on the potential for its medical application, it is by attending to the social life of DNA that we can appreciate—and truly assess—our collective condition. Reconciliation projects spurred by DNA testing may be starting points for such dialogues, but we cannot rely on science to propel social change.

When I began my research into genetic genealogy over a decade ago, I could not have imagined that I would end up writing a book about racial politics in the United States. These were small stories, personal accounts that, while often deeply felt, did not rise to the level of large-scale social issues. But what I found, of course, ran the gamut from the performance of identity to the seeking of justice.

In this journey, there were so many people along the way who helped me to both see and do the work of telling the bittersweet story of how the long struggle of racial equality propelled tactical ingenuity—the use of cutting-edge technologies immediately put to unintended purposes following their release—but had strategic constraints, as new technologies often hold for black communities.

In looking at the uses and abuses of genetic claims, one can only conclude that DNA is Janus-faced. Yet with genetic ancestry testing, these two ways of looking are also sankofa. The sankofa is an Adinkra symbol from the Akan community of Ghana. When the burials were uncovered at the African Burial Ground, one of the coffins had a heart-shaped symbol made of iron nails tacked into its top. This symbol, which some scholars believe to be the sankofa, was incorporated into the African Burial Ground Monument as the most prominent of a set of symbols. A rendering of the design found on the coffin lid is prominently carved into a large black granite memorial at the center of the site and serves as a symbol for the African Burial Ground as a whole. The meaning of the sankofa symbol has been differently interpreted, but those interpretations aptly converge around going back or looking back to something or some time forgotten.

The boom in genetic ancestry testing over the last decade has been extraordinary. It's ever-rising and decade-plus of staying power confirms that this pursuit is neither a fad nor a trend. For good and for naught, we use DNA as a portal to the past that yields insights for the present and the future. We use DNA to shine a light on social trauma and to show how historic injustices continue to resonate today.

Under social conditions in which discrimination and injustice experienced by African Americans both in the past and in the present are discounted or denied, the turn to forensic evidence as a way to compel recognition is understandable. Just as we hope that body cameras might serve as a check on law enforcement abuses, we seek genetic genealogy to expose and substantiate the legacies of racial slavery. Given black communities' accumulated mistrust of science, the use of genetic analysis for liberatory ends is perhaps paradoxical. But the transformative aspirations for DNA are consistent with the long-standing use of empirical data in racial justice struggles, from the "doll studies" conducted by psychologists Kenneth Clark and Mamie Phipps Clark that would serve as "exhibit A" in the *Brown v. Board of Education* decision of 1954 to the Mapping Police Violence project that collects aggregate data about men and women killed by law enforcement.[3] But as the field of agnotology—the study of culturally induced ignorance—shows us, evidence may be no match for ideology. Genetic ancestry testing is but one implement in an entire tool kit of tactics that, marshaled together, must be brought to the project of building racial reconcilation and social justice.

ACKNOWLEDGMENTS

In the early days of this project, remarkably generous members of the Afro-American Historical and Genealogical Society, particularly the Jean Sampson Scott Greater New York and Central California chapters of the organization, showed me the family-history ropes with patience and wisdom. I especially thank Dr. Stanton Biddle, Doris Burbridge, Denise Lancaster-Young, Andrea Butler Ramsey, Alene Jackson Smith, Adeline Jackson Tucker, and Sharon Wilkins for their many kindnesses.

I owe a debt of gratitude to the scores of genealogists, social scientists, and scientists who took time to speak with me, many with the promise of anonymity. I thank them all for their generosity, including Jane Aldrich, Blaine Bettinger, Michael Blakey, Henry Louis Gates Jr., Alan Goodman, Robert L. Hall, William Holland, Fatimah Jackson, Rick A. Kittles, Beaula McCalla, Joseph Opala, Gina Paige, Warren Perry, Chandler Saint, Gregory Simpkins, Linda D. Strausbaugh, Arthur Torrington, CBE, and Isaiah Washington. My appreciation also goes to the descendants of Venture and Margaret Smith, especially the late Coralynne Henry Jackson and Florence Warmsley, who generously shared their thoughts about their family with me.

Audiences at the Bates College, the Brocher Foundation, Brown University, City University of New York–Graduate Center, Columbia University, Cornell University, Duke University, the Max Planck

Institute for the History of Science, MIT, the New York Institute of the Humanities, New York University, Princeton University, Rensselaer Polytechnic, Rutgers University, the Smithsonian National Museum of Natural History, Temple University, Yale University, the University of California at Los Angeles, the University of California at Santa Cruz, the University of Michigan, the University of Pennsylvania, and the University of Toronto offered valued feedback at critical states of the project's evolution.

Stephanie Alvarado, Mary Barr, Felicia Bevel, Christina Ciocca, Milo Inglehart, Tamika Jackson, Julia Mendoza, Thalia Sutton, and Shondrea Thornton ably assisted me with research. Thank you Ifeoma Ajunwa, Althea Anderson, Cecilia Cardenas-Navia, Sean Greene, Stephanie Greenlea, Tisha Hooks, Nicole Ivy, James Jones, Elyakim Kislev, Warren McKinney, Ronna Popkin, James T. Roane, Joan H. Robinson, Daniel Royles, Anthony Ureña, Emily Vasquez, and Devon Wade for helping me to sharpen my thinking as the book was in progress.

Generous colleagues read or discussed the manuscript at various stages and gave me valuable feedback, including Elizabeth Alexander, Edward Ball, Ruha Benjamin, Catherine Bliss, danah boyd, Adele Clarke, Kate Crawford, Steven Epstein, Joan Fujimura, Duana Fullwiley, Paul Gilroy, Jennifer Hamilton, Ange-Marie Hancock, Saidiya Hartman, Stefan Helmreich, Marianne Hirsch, Jonathan Holloway, Gerald Jaynes, Kellie Jones, Randall Kennedy, Shamus Khan, Jennifer Lena, Delores Y. Nelson, Tamara Nopper, Anne Pollock, Ramya Rajangopalan, Dorothy Roberts, Salamishah Tillet, Kendall Thomas, Keith Wailoo, Vron Ware, Patricia J. Williams, Audra Wolfe, and Eviatar Zerubavel. My incredible writing group, which at times spanned four time zones, always helped me to see what the project could be; exuberant thanks to Catherine Lee, Ann Morning, and Wendy Roth. Jenny Reardon is one of my most valued interlocutors, and our regular conversations always left me smarter.

The Renaissance man, Thomas Sayers Ellis, took my author's photo and I thank him for sharing the gift of his magical eye.

I am grateful to my agent, Deirdre Mullane of Mullane Literary, who immediately saw the possibilities for this project and brought it to the attention of the historic Beacon Press. Gayatri Patnaik, my visionary and discerning editor at Beacon, brought clarity and focus to the manuscript, time and time again. Her colleague Rachael Marks provided insightful comments about the text and was a great steward of it. Melissa Dobson was a sterling copy editor.

My thanks to dear colleagues and friends Nadia Abu El-Haj, Jafari Allen, Karen Barkey, Peter Bearman, Marcellus Blount, George Chauncey, Peter Chow-White, Jelani Cobb, Yinon Cohen, Dalton Conley, Jessie Daniels, Thomas DiPrete, Troy Duster, Gil Eyal, Farah Ron Gregg, Jasmine Griffin, Evelynn Hammonds, Fredrick Harris, Carl Hart, Tayari Jones, Paul D. Miller, Lisa Nakamura, Mark Anthony Neal, Aaron Panofsky, Elizabeth Povinelli, Guthrie Ramsey, Adam Reich, Susan Reverby, Charmaine Royal, Saskia Sassen, Carla Shedd, Rebecca Skloot, Sy Spillerman, David Stark, Thomas Thurston, Van Tran, Vina Tran, Thuy Linh Tu, Diane Vaughn, Sudhir Venkatesh, Harriet Washington, Josh Whitford, and Isabel Wilkerson.

My most heartfelt appreciation to my extended family—the Nelsons, the Mundys, the Williamses, the Alexanders, the Ghebreyesuses, and the Etiennes. I am happily beholden to Andrea Nelson Saunders, Aaron Saunders, Robert Nelson Jr., Dawn Nelson, Anthony Nelson, and Vera Nelson for their love and support. I am continually awed by the boundless promise of Aidan Nelson, Austin Nelson, Anthony Nelson Jr., Alexis Nelson, Alondra Christine Hall, Anita Hall, Ariella Nelson, Brianna Nelson, Bryce Saunders, Joseph Hall Jr., Mya Nelson, Reina Saunders, and Renee Nelson. I am proud to call Joseph Kim and Jonathan Merino kin. My great-nieces, Valentina, Ava, and Camilla, delight me to no end. My loving parents, Robert S. Nelson Sr. and Delores Y. Nelson, always believe I am better than I am. My love, Garraud Etienne, is simply the best.

NOTES

INTRODUCTION

1. Linda Strausbaugh et al., "The Genomics Perspective on Venture Smith: Genetics, Ancestry, and the Meaning of Family," in *Venture Smith and the Business of Slavery and Freedom*, ed. James Brewer Stewart (Amherst: University of Massachusetts Press, 2010), 209.

2. James Brewer Stewart, editor's preface to *Venture Smith and the Business of Slavery and Freedom*, xiv.

3. Venture Smith, *A Narrative of the Life and Adventures of Venture, a Native of Africa, but Resident Above Sixty Years in the United States of America, Related by Himself* (New London, CT: Printed by C. Holt, at the Bee-office, 1789). In the account, Smith explains that his first owner, James Mumford, called him Venture because he was purchased through Mumford's "own private venture." With his last owner, Oliver Smith, he entered into an arrangement whereby he worked extra jobs in order to buy his freedom.

4. Luke Collingwood was the notorious mastermind of a 1781 massacre during which more than a hundred enslaved men and women were murdered. Collingwood and his human cargo of more than four hundred set out for Jamaica from Africa on a vessel named the *Zong*. When the vessel ran off course, extending an already perilous journey and risking the health of the captives for whom his patrons would be financially liable should they die, Collingwood ordered them thrown off the vessel into the sea so that insurance damages could be claimed.

5. Stewart, editor's preface, *Venture Smith and the Business of Slavery and Freedom*, xiv.

6. Strausbaugh et al., "The Genomics Perspective on Venture Smith," 208.

7. Ibid., 225.

8. Dorothy Nelkin and M. Susan Lindee, *The DNA Mystique: The Gene as a Cultural Icon*, 2nd ed. (Ann Arbor: University of Michigan Press, 2004), xxix.

9. Ibid., 2.

10. There are now more than three dozen genetic-ancestry-testing companies. Last summer, 23andMe announced that it had genotyped one million individuals. See Anne Wojcicki, "Power of One Million," *23andMe Blog*, June 18, 2005, http://blog.23andme.com/news/one-in-a-million/. A 2014 story in the *Scientist* placed Ancestry.com and Family Tree DNA's clients at 600,000 each. See Tracy Vence, "DNA Ancestry for All," *Scientist*, July 10, 2014, http://www.the-scientist.com/?articles.view/articleNo/40460/title/DNA-Ancestry-for-All/. In a 2015 interview with the author, Gina Paige of African Ancestry numbered that company's niche market of customers of African descent at 45,000 over the last twelve years.

11. Herbert J. Gans (1979): "Symbolic Ethnicity: The Future of Ethnic Groups and Cultures in America," *Ethnic and Racial Studies* 2, no. 1 (1979): 1–20. A helpful chronology of the "Americanization" of root-seeking from the colonial era to the present is provided in Francois Weil, *Family Trees: A History of Genealogy in America* (Cambridge, MA: Harvard University Press, 2013).

12. *A Slave's Story*, British Broadcasting Company, March 25, 2007.

13. Ibid. Researchers obtained several samples from Smith's coffin, but no DNA was recovered.

14. Nelkin and Lindee, *The DNA Mystique*, xii.

15. See Nicholas Wade, "A Decade Later, Genetic Map Yields Few New Cures," *New York Times*, June 10, 2010, www.nytimes.com/2010/06/13/health/research/13genome.html; "Spiegel Interview with Craig Venter: 'We Have Learned Nothing from the Genome,'" *Spiegel Online International*, July 29, 2010, http://www.spiegel.de/international/world/spiegel-interview-with-craig-venter-we-have-learned-nothing-from-the-genome-a-709174.html.

16. Arjun Appadurai, *The Social Life of Things: Commodities in Cultural Perspective* (New York: Cambridge University Press, 1986), 5.

17. Alondra Nelson, "The Social Life of DNA," *Chronicle of Higher Education*, August 29, 2010, http://chronicle.com/article/The-Social-Life-of-DNA/124138/.

18. Alondra Nelson, "Reconciliation Projects: From Kinship to Justice," in *Genetics and the Unsettled Past: The Collision of DNA, Race, and History*, ed. Keith Wailoo, Alondra Nelson, and Catherine Lee (Rutgers, NJ: Rutgers University Press, 2012), 20–31.

19. Christine Hine, *Virtual Ethnography* (London: Sage Publications, 2000); Daniel Miller and Don Slater, *The Internet: An Ethnographic Approach* (Oxford, UK: Berg, 2001); Stefan Helmreich, "Spatializing Technoscience," *Reviews in Anthropology* 32 (2003): 13–36. See also Deborah Heath et al., "Nodes

and Queries: Linking Locations in Networked Fields of Inquiry," *American Behavioral Scientist* 43 (1999): 450–63.

20. A. J. Hostetler, "Who's Your Daddy? Genealogists Look Inside Their Cells for Clues to Their Ancestors," *Richmond (VA) Times-Dispatch*, April 24, 2003; Stephen Magagnini, "DNA Helps Unscramble the Puzzles of Ancestry," *Sacramento Bee*, August 3, 2003; Steve Sailer, "African Ancestry, Inc., Traces DNA Roots," *Washington Times*, April 28, 2003; Frank D. Roylance, "Reclaiming Heritage Lost to Slavery," *Baltimore Sun*, April 17, 2003; Rick A. Kittles, interview with author, February 4, 2006.

21. Sam Fulwood III, "His DNA Promise Doesn't Deliver," *Los Angeles Times*, May 29, 2000, http://articles.latimes.com/2000/may/29/news/mn-35219.

22. Ibid.

23. Barbara Katz Rothman, *Genetic Maps and Human Imaginations: The Limits of Science in Understanding Who We Are* (New York: Norton, 1998); Troy Duster, *Backdoor to Eugenics* (New York: Routledge, 1990); and Nelkin and Lindee, *The DNA Mystique.*

24. Donald Kennedy, "Not Wicked Perhaps, but Tacky," *Science* 297 (2002): 1237; Craig J. Venter, "A Part of the Human Genome Sequence," *Science* 299 (2003): 1183–84.

25. Craig J. Venter et al., "The Sequence of the Human Genome," *Science* 291 (2001): 1304–51.

26. "Reading the Book of Life: White House Remarks on Decoding of Genome," *New York Times*, June 27, 2000, http://www.nytimes.com/2000/06/27/science/reading-the-book-of-life-white-house-remarks-on-decoding-of-genome.html.

27. Dorothy Roberts, *Fatal Invention: How Science, Politics and Big Business Re-Create Race in the Twenty-First Century* (New York: New Press, 2011), x.

28. Ibid.

29. Vincent Sarich and Frank Miele, *Race: The Reality of Human Differences* (Boulder, CO: Westview Press, 2004).

30. Neil Risch et al., "Categorizations of Humans in Biomedical Research: Genes, Race, and Disease," *Genome Biology* 3, no. 7 (2002): Comment 2007.1–2007.12; Esteban G. Burchard et al., "The Importance of Race and Ethnic Background in Biomedical Research and Clinical Practice," *New England Journal of Medicine* 348 (2003): 1174.

31. Nicholas Wade, *A Troublesome Inheritance: Genes, Race, and Human History* (New York: Penguin, 2014).

32. See Eduardo Bonilla-Silva, *Racism Without Racists: Color-Blind Racism and the Persistence of Racial Inequality in the United States* (Lanham, MD: Rowman & Littlefield, 2003).

33. Roberts, *Fatal Invention*, 287.

34. Stephanie Greenlea, "Free the Jena Six! Racism, Technology and Black

Solidarity in the Digital Age" (PhD diss., Yale University, 2012), 4. See also Bonilla-Silva, *Racism Without Racists.*

35. On the need to bring racism into awareness with respect to twenty-first-century social movements, see ibid.

36. Dena S. Davis, "Genetic Research & Communal Narratives," *Hastings Center Report* 34, no. 4 (July–August 2004): 43.

37. Natalie Angier, "Scientist at Work: Mary-Claire King; Quest for Genes and Lost Children," *New York Times*, April 23, 1993, http://www.nytimes.com/1993/04/27/science/scientist-at-work-mary-claire-king-quest-for-genes-and-lost-children.html?pagewanted=all&src=pm. This reconciliation project is discussed at length in chapter 1.

38. Peter Gill et al., "Identification of the Remains of the Romanov Family by DNA Analysis," *Nature Genetics* 6, no. 2 (February 1994): 130–36.

39. Molecular anthropologist Frederika Kaestle, an expert in the analysis of ancient DNA, was tasked with carrying out analysis of the remains. (Ancient DNA techniques are capable of being used on samples that are up to one hundred thousand years old.) Study of the remains was halted after local indigenous groups invoked their rights under NAGPRA, but researchers sued the federal government to regain access. The judge hearing the suit requested that preliminary investigation of the remains be done so that their research potential could be ascertained. Subsequent DNA analysis took place in 2000, by a team of researchers lead by Kaestle. But these remains had begun to fossilize and little information could be extracted from them. See Frederika Kaestle, "Report on DNA Analysis of the Remains of 'Kennewick Man' from Columbia Park, Washington," in F. P. McManamon, *Kennewick Man* (Washington, DC: US Department of the Interior, National Park Service, May 2004), http://www.nps.gov/archeology/kennewick/Kaestle.htm. For an excellent critique of the use of genetics in the "Kennewick Man" case, see also Kimberly TallBear, "Genetics, Identity and Culture in Indian Country" (working paper, International Institute for Indigenous Resource Management, Denver, CO, 2000), and Michelle M. Jacob, "Making Sense of Genetics, Culture, and History: A Case Study of a Native Youth Education Program," in Wailoo, Nelson, and Lee, *Genetics and the Unsettled Past*, 279–94.

40. Morten Rasmussen et al., "The Ancestry and Affiliations of Kennewick Man," *Nature* (2015), doi: 10.1038/nature14625.

41. Annette Gordon-Reed, *Thomas Jefferson & Sally Hemings: An American Controversy* (Charlottesville: University of Virginia Press, 1997); Annette Gordon-Reed, *The Hemingses of Monticello: An American Family* (New York: Norton, 2008).

42. Eugene A. Foster et al., "Jefferson Fathered Slave's Last Child," *Nature* 396 (1998): 27–28. In 1999 the scientists would qualify their findings, saying that the father of Hemings's last son could have been Thomas Jefferson or one of a small number of his paternal relatives.

43. Ibid. The rarity of the haplotype shared by paternal male Jefferson descendants was subsequently reconfirmed. See Turi E. King et al., "Thomas Jefferson's Y Chromosome Belongs to a Rare European Lineage," *American Journal of Physical Anthropology* 132 (2007): 584–89.

44. John Works, "The Jefferson-Hemings Controversy: A New Critical Look," *Drumbeat* (Fall 2010): 17–19, http://www.tjheritage.org/newscomfiles/WorksJefferson-HemingsArticle.pdf.

45. Jackson founded the African-American DNA Roots Project, a not-for-profit genetic ancestry analysis service, with Bert Ely, a biologist at the University of South Carolina, in 2001.

46. Jonathan Mummolo, "African American Seeks to Prove Genetic Link to James Madison," *Washington Post*, June 11, 2007, http://www.washingtonpost.com/wp-dyn/content/article/2007/06/10/AR2007061001404.html; "Proving Lineage to a President," National Public Radio, June 14, 2007, http://www.npr.org/templates/story.php ?storyId=11077716; Kris Coronado, "What Ever Happened To . . . the Possible Relative of James Madison," *Washington Post*, October 14, 2011, http://www.washingtonpost.com/lifestyle/magazine/whatever-happened-to--the-possible-relative-of-james-madison/2011/09/27/gIQA9evGkL_story.html.

47. Sierra Express Media, "Isaiah Washington States His Sierra Leone Passport Has No Bearing on What He Can Do for Salone," *Sierra Express*, September 18, 2014, http://www.sierraexpressmedia.com/?p=70465; "Isaiah Washington's Fight Against Ebola," CNN, April 17, 2015, http://www.cnn.com/videos/world/2015/04/17/ebola-ball-challenge-isaiah-washington-intv.cnn.

CHAPTER 1: RECONCILIATION PROJECTS

1. John Torpey, *Making Whole What Has Been Smashed: On Reparations Politics* (Cambridge, MA: Harvard University Press, 2006).

2. King and a group of collaborators, including Cavalli-Sforza, also played an important role in the evolution of the genomics era, issuing a call, in the journal *Genomics*, for an extension of the Human Genome Project that would consist of a "worldwide survey of human genetic diversity." This well-intentioned, visionary but ultimately flawed idea, contributed to the creation of the parallel Human Genome Diversity Project. See Luca Cavalli-Sforza et al., "Call for a Worldwide Survey of Human Genetic Diversity: A Vanishing Opportunity for the Human Genome Project," *Genomics* 11 (1991): 490–91.

3. David Noonan, "The Genes of War," *Discover*, October 1990, 50.

4. Ibid., 51–52; Lisa Yount, "King, Mary-Claire," in *A to Z of Women in Science and Math* (New York: Facts on File, 2007).

5. Allan C. Wilson et al., "Mitochondrial DNA and Two Perspectives on Evolutionary Genetics," *Biological Journal of the Linnean Society of London* 26, no. 4 (1985): 375–400.

6. Noonan, "The Genes of War," 52.

7. Rory Carroll and Jeff Farrell, "Argentina's Authorities Order DNA Tests in Search for Stolen Babies of Dirty War," *Guardian*, December 30, 2009, http://www.theguardian.com/world/2009/dec/30/argentina-dna-tests-babies-disappeared.

8. Julia Kumari Drapkin, "Torn Between Identities in Argentina," *Global Post*, November 11, 2010, http://www.globalpost.com/dispatch/argentina/101103/dna-clarin-dirty-war.

9. Noonan, "The Genes of War," 51.

10. Linsday Adams Smith, "Subversive Genes: Re(con)stituting Identity, Family and Human Rights in Argentina" (PhD diss., Harvard University, 2009). See also Lindsay A. Smith and Sarah Wagner, "DNA Identification," *Anthropology News* (May 2007): 35.

11. Quoted in Catherine Bliss, *Race Decoded: The Genomic Fight for Social Justice* (Stanford, CA: Stanford University Press, 2012), 75.

12. Rick Kittles, "Tracing Our Ancestors," Chicago Humanities Festival, November 9, 2013.

13. Duana Fullwiley, "Race, Genes, Power," *British Journal of Sociology* 66 (2015): 43.

14. Melissa Nobles, *The Politics of Official Apologies* (New York: Cambridge University Press, 2008), 3, 40.

15. Greensboro Truth and Reconciliation Commission, "November 3, 1979," http://greensborotrc.org/november3.php.

16. Greensboro Truth and Reconciliation Commission, "The Mandate and Guiding Principles," http://greensborotrc.org/mandate.php.

17. Greensboro Truth and Reconciliation Commission Report, Executive Summary, May 25, 2006, http://greensborotrc.org/exec_summary.pdf.

18. John Torpey, ed., *Politics and the Past: On Repairing Historical Injustices* (Lanham, MD: Rowan and Littlefield, 2003), 1.

19. Michael Fortun, *Promising Genomics: Iceland and deCODE Genetics in a World of Speculation* (Berkeley: University of California Press, 2008), 10.

20. Susan Dwyer, "Reconciliation for Realists," *Ethics and International Affairs* 13 (1999): 96.

CHAPTER 2: GROUND WORK

1. American Antiquities Act of 1906, Section 2, http://www.nps.gov/history/local-law/anti1906.htm.

2. Erika Hagelberg, Bryan Sykes, and Robert Hedges, "Ancient Bone DNA Amplified," *Nature* 342 (November 1989): 485. The African Burial Ground researchers carried out their first, partially successful, genetic analysis of the burial remains in 1995. By temporal comparison, DNA testing resolving the Jefferson-Hemings controversy was conducted in 1998 and genetic analysis of the Kennewick Man in Washington State was conducted in 2000.

3. Spencer P. M. Harrington, "Bones and Bureaucrats," *Archaeology* (March/April 1993), http://www.archaeology.org/online/features/afrburial/.

4. Quoted in ibid.

5. David W. Dunlap, "Excavation Stirs Debate on Cemetery," *New York Times*, December 6, 1991, http://www.nytimes.com/1991/12/06/nyregion/excavation-stirs-debate-on-cemetery.html.

6. Ibid.

7. Paterson, quoted in ibid.

8. Cheryl J. La Roche and Michael L. Blakey, "Seizing Intellectual Power: The Dialogue at the New York African Burial Ground," *Historical Archaeology* 31 (1997): 84–106.

9. Dunlap, "Excavation Stirs Debate on Cemetery."

10. "Paterson to Monitor Dig at Burial Ground," *New York Times*, December 7, 1991, http://www.nytimes.com/1991/12/07/nyregion/paterson-to-monitor-dig-at-burial-ground.html.

11. Warren R. Perry, "Archaeology as Community Service: The African Burial Ground Project in New York City" (University of South Florida, St. Petersburg, n.d.), http://www.stpt.usf.edu/~jsokolov/burialgr.htm, accessed September 29, 2010. Interview with author, July 12, 2012.

12. Activity resumed briefly in September 1992 in order to complete the excavation of "ten burials and one grave pit" that had been left partly exposed after work was halted two months prior.

13. Michael L. Blakey, "The New York African Burial Ground Project: An Examination of Enslaved Lives, a Construction of Ancestral Ties," *Transforming Archaeology* 7, no. 1 (1998): 53–58; Felicia R. Lee, "Harlem's Cultural Anchor in a Sea of Ideas," *New York Times*, May 11, 2007, http://www.nytimes.com/2007/05/11/arts/design/11harl.html.

14. Glen C. Campbell, AIA, ISP (GC), to the General Services Administration of the New York African Burial Ground Peer Review Panel, Schomburg Center Archive, p. 1.

15. Michele N-K Collison, "Disrespecting the Dead," *Black Issues in Higher Education* (April 1, 1999): 16–17. On funding, see Blakey, "The New York African Burial Ground Project."

16. Michael L. Blakey, PhD, to Ms. Gina L. Stahlnecker, district assistant to Senator David A. Paterson, April 9, 1992, Schomburg Center for Research in Black Culture Archive.

17. For more about debates on race in forensic osteology, see Marc R. Feldesman and Robert L. Fountain, "'Race' Specificity and the Femur/Stature Ratio," *American Journal of Physical Anthropology* 100 (June 1996): 207–24; and Stephen Ousley, Richard Jantz, and Donna Freid, "Understanding Race and Human Variation: Why Forensic Anthropologists Are Good at Identifying Race," *American Journal of Physical Anthropology* 139 (January 2009): 68–76.

18. As well as archaeologists, paleoarchaeologists, biological anthro-

pologists, and forensic scientists. See Michael L. Blakey, PhD, to Dr. James V. Taylor, director, Metropolitan Forensic Anthropology Team, Lehman College, CUNY, December 1, 1992, 1–2, Schomburg Center for Research in Black Culture Archive.

19. La Roche and Blakey, "Seizing Intellectual Power," 88. See also Blakey, introduction to "The New York African Burial Ground Project."

20. Ibid.

21. La Roche and Blakey, "Seizing Intellectual Power," 88.

22. Blakey, "The New York African Burial Ground Project," 53.

23. George J. Armelagos and Dennis P. Van Gerven, "A Century of Skeletal Biology and Paleopathology: Contrasts, Contradictions, and Conflicts," *American Anthropologist* 105, no. 1 (2003): 53.

24. La Roche and Blakey, "Seizing Intellectual Power," 88. This interpretation of the MFAT investigators' intentions was supported by the research design submitted by HCI in advance of the burial ground survey. The complete proposal comprised just twelve pages detailing the removal and storage of the remains. Very little of the HCI's course of action was devoted to research on the burials.

25. Gordon R. Mitchell and Kelly Happe, "Defining the Subject of Consent in DNA Research," *Journal of Medical Humanities* 22 (2001): 49.

26. Michael L. Blakey, PhD, to Dr. James V. Taylor, director, Metropolitan Forensic Anthropology Team, Lehman College, CUNY, December 1, 1992, 1–2, Schomburg Center for Research in Black Culture Archive.

27. Stephen Jay Gould, *The Mismeasure of Man* (New York: Norton, 1996 [1981]), 100–101. On research into race and lung capacity, see Lundy Braun, "Spirometry, Measurement, and Race in the Nineteenth Century," *Journal of the History of Medicine and Allied Sciences* 60, no. 2 (2005): 135–69.

28. Michael A. Gomez, *Exchanging Our Country Marks: The Transformation of African Identities in the Colonial and Antebellum South* (Chapel Hill: University of North Carolina Press, 1998), 154. There is a significant body of research that takes up the question of the convergence and divergence between concepts of race and ethnicity in the United States, and I recognize the interdependence of these terms. In drawing a distinction between them here, I aim to highlight the particular stakes of the African Burial Ground controversy.

29. Ibid., 12.

30. Blakey, quoted in Mitchell and Happe, "Defining the Subject of Consent in DNA Research," 48.

31. La Roche and Blakey, "Seizing Intellectual Power," 86.

32. This difference in research interpretation also reflected a wider scholarly debate over whether archaeology was best conducted through a forensic framework or an anthropological one. See, for example, Sherwood L. Washburn. "The New Physical Anthropology," *Transactions of the New York Academy of Science*, series 2, 13 (1951): 298–304; and Diana B. Smay and George

J. Armelagos, "Galileo Wept: A Critical Assessment of the Use of Race in Forensic Anthropology," *Transforming Anthropology* 9 (2000): 19–40.

33. F. L. C. Jackson et al., "Origins of the New York Burial Ground Population: Biological Evidence of Geographical and Macroethnic Affiliations Using Craniometrics, Dental Morphology, and Preliminary Genetic Analysis," in *The New York African Burial Ground: History Final Report*, ed. Edna Greene Medford (Washington, DC: US General Services Administration, November 2004), 153–54. See also Blakey's introduction, *The New York African Burial Ground*, 17–22.

34. Jackson et al., "Origins of the New York Burial Ground Population," 150.

CHAPTER 3: GAME CHANGER

1. Interview with author, July 25, 2012.

2. Ibid.

3. Jackson et al., "Origins of the New York Burial Ground Population," 150.

4. The large research group included Blakey, University of Maryland anthropologist Fatimah Jackson, Hampshire College anthropologist Alan Goodman, Boston University historian Linda Heywood, Howard University geneticist Matthew George, physical anthropologists Warren Perry and S.O.Y. Keita, and Rick Kittles, a doctoral student in biology from George Washington University.

5. Jackson et al., "Origins of the New York Burial Ground Population," 150.

6. Michael Blakey, "Bioarchaeology of the African Diaspora in the Americas," *Annual Review of Anthropology* (2001).

7. Nancy Leys Stepan and Sander L. Gilman, "Appropriating the Idioms of Science: The Rejection of Scientific Racism," in *The "Racial" Economy of Science: Toward a Democratic Future*, ed. Sandra Harding (Bloomington: Indiana University Press, 1993), 170.

8. Jackson et al., "Origins of the New York African Burial Ground Population," 69.

9. *African Burial Ground* (Final Report), chapters 5, 11, and 13.

10. Jackson et al., "Origins of the New York African Burial Ground Population," 150.

11. Interview with author, June 14, 2012.

12. Jackson et al., "Origins of the New York African Burial Ground Population," 151.

13. Ibid., 162.

14. Ibid. Also Rick Kittles, "Inferring African Ancestry of African Americans," Annual Meeting of the American Association of Physical Anthropologists, March 30, 2001 (unpublished paper).

15. Michael L. Blakey, PhD, "Research Design for Temporary Curation and Anthropological Analysis of the 'Negro Burying Ground' (Foley Square) Archaeological Population at Howard University," June 11, 1992, p. 2, Schomburg Center for Research in Black Culture Archive.

16. "Report to the General Services Administration of the New York African Burial Ground Peer Review Panel, May 12–14, 1993," p. 6, Schomburg Center Archive.

17. Fatimah Jackson, "Concerns and Priorities in Genetic Studies: Insights from Recent African American Biohistory," *Seton Hall Law Review* 27 (1997): 957.

18. "Report to the General Services Administration of the New York African Burial Ground Peer Review Panel," 17.

19. Ibid.

20. Blakey, "Research Design for Temporary Curation and Anthropological Analysis of the 'Negro Burying Ground,'" 15. See also Allan C. Wilson et al., "Mitochondrial DNA and Two Perspectives on Evolutionary Genetics," *Biological Journal of the Linnean Society of London* 26, no. 4 (1985): 375–400.

21. Ibid.

22. Alan Cooper and Hendrik N. Poinar, "Ancient DNA: Do It Right or Not at All," *Science* 289 (2000): 1139.

23. Jackson et al., "Origins of the New York African Burial Ground Population," 85.

24. John Simons, "Out of Africa," *Fortune*, February 19, 2007, p. 37.

25. Jackson et al. "Origins of the New York African Burial Ground Population," 86.

26. "Return to the African Burial Ground: An Interview with Physical Anthropologist Michael L. Blakey," *Archaeology*, November 20, 2003, http://www.archaeology.org/online/interviews/blakey/index.html.

27. Mark A. Jobling and Chris Tyler-Smith, "Fathers and Sons: The Y Chromosome and Human Evolution," *Trends in Genetics* 11 (November 1995): 449–56.

28. Jackson et al., "Origins of the New York African Burial Ground Population," 88.

29. Simons, "Out of Africa," 37.

30. Jackson et al., "Origins of the New York African Burial Ground Population," 87.

31. Ibid., 89, 91.

32. Blakey, "The New York African Burial Ground Project," 55–56. He continues, "Muscle attachments become enlarged when muscles undergo frequent strain. Most of the population of men and women have [*sic*] enlarged muscle attachments in the neck, arms, and legs. It has become commonplace for our technicians to find women's bones that are so robust as to be indistinguishable from those of men." See also Harrington, "Bones and Bureaucrats."

33. David Willard, "Dignity for the Living," *William and Mary News*, December 12, 2003, https://www.wm.edu/news/stories/2003/dignity-for-the-living.php.

34. White House, "Establishment of the African Burial Ground National Monument," press release, February 27, 2006, http://georgewbush-whitehouse.archives.gov/news/releases/2006/02/20060227-6.html.

35. Harrington, "Bones and Bureaucrats."

36. Ibid.

37. Carey Goldberg, "DNA Offers Link to Black History," *New York Times*, August 28, 2000, http://www.nytimes.com/2000/08/28/us/dna-offers-link-to-black-history.html.

38. Deborah Bolnick et al., "The Business and Science of Ancestry Testing," *Science* 318, no. 5849 (2007): 399–400.

39. Goldberg, "DNA Offers Link to Black History." See also Jackson et al., "Origins of the New York African Burial Ground Population," 151.

40. Quoted in "A Year After Reburial of Slaves, Debate over Memorial," Associated Press, October 21, 2004, in *Diverse*, http://diverseeducation.com/article/4063/. See also Saeed Shabazz, "Howard U Scientists Make Historic DNA Breakthrough," *Final Call*, January 1, 2000, http://www.finalcall.com/artman/publish/Modern_Technology_15/Howard_U_scientists_make_historic_DNA_breakthrough_4017.shtml.

41. Blakey, "The New York African Burial Ground Project," 55.

42. Author interview with Warren Perry, July 12, 2012.

43. Fatimah Jackson, "Concerns and Priorities in Genetic Studies: Insights from Recent African American Biohistory," *Seton Hall Law Review* 951 (1996–97): 957.

44. Ibid.

45. Interview with author, 2015.

46. Interview with author, 2014.

47. Ayana Jones, "African Ancestry Come to Life," *Philadelphia Tribune*, May 22, 2012, http://www.phillytrib.com/businessarticles/item/4195-african-ancestry-comes-to-life.html.

48. Established in 2001, Sorenson was one of the first US companies to specialize in high-efficiency DNA sequencing and genotyping. It conducts analysis for population genetics as well as DNA genealogy, forensic genetics, and medical genetics. The company is also in the business of DTC testing. In 2007 it acquired Identigene, which provides DNA paternity tests that can be purchased in drug stores and at other locations.

49. African Ancestry, African Lineage Database, http://www.africanancestry.com/database.html.

50. African Ancestry reports that approximately 25 to 30 percent of male root-seekers using its PatriClan (Y chromosome) test will not match any of the paternal lines in the African Lineage Database (ALD). In such instances,

the customer may be advised to have his sample matched against a "European database." See Greg Langley, "Genealogy and Genomes: DNA Technology Helping People Learn More About Who They Are and Where They Come From," *Baton Rouge Advocate*, July 20, 2003. A page about the PatriClan (Y chromosome) test on African Ancestry's website states: "We find African ancestry for approximately 65% of the paternal lineages we test. The remaining 35% of the lineages we test typically indicate European ancestry. If our tests indicate that you are not of African descent, we will identify your continent of origin." "Discover the Paternal Roots of Your Family Tree," http://africanancestry.com/patriclan.html (accessed July 1, 2010). Because the ALD is extensive but not exhaustive, however, it is possible that some African genetic markers are not yet included.

51. Alondra Nelson, "The Factness of Diaspora," in *Revisiting Race in a Genomic Age*, ed. Barbara A. Koenig, Sandra Soo-Jin Lee, and Sarah S. Richardson (New Brunswick, NJ: Rutgers University Press, 2008), 255; Alondra Nelson, "Reconciliation Projects: From Kinship to Justice," in Wailoo, Nelson, and Lee, *Genetics and the Unsettled Past*.

52. African Burial Ground Foundation Memorial Dedication postcard, September 2007.

53. James D. Faubion and Jennifer Hamilton, "Sumptuary Kinship," *Anthropological Quarterly* 80 (2007): 540.

CHAPTER 4: THE PURSUIT OF AFRICAN ANCESTRY

1. Alex Haley, *Roots: The Saga of an American Family* (New York: Dell, 1976).

2. Helen Taylor, "'The Griot from Tennessee': The Saga of Alex Haley's *Roots*," *Critical Quarterly* 37 (June 1995): 47.

3. Ibid.

4. Ibid., 46–47.

5. Philip Nobile, "Uncovering Roots," *Village Voice*, February 23, 1993.

6. Taylor, "'The Griot from Tennessee,'" 48.

7. Carolyn J. Rosenthal, "Kinkeeping in the Familial Division of Labor," *Journal of Marriage and the Family* 47 (1985): 965–74.

8. Genetic genealogy testing is employed by other diasporic social groups whose historical experiences of migration, dispersal, and persecution have made it difficult to document genealogical information, including the Irish (Catherine Nash, "Genetic Kinship," *Cultural Studies* 18 [2004]: 1–33) and Jewish (Nadia Abu El-Haj, "'A Tool to Recover Past Histories': Genealogy and Identity After the Genome," Occasional Paper No. 19 [December 2004], School of Social Science, Institute for Advanced Study, Princeton, NJ, www.sss.ias.edu/publications/papers/paper19.pdf). These services are also widely used for religious reasons. For example, genealogy is an important part of the after-life cosmology of the Church of Jesus Christ of Latter-day Saints.

9. Quotes in Taylor, "'The Griot from Tennessee,'" 49.

10. Maurice Halbwachs, *On Collective Memory* (Chicago: University of Chicago Press, 1992).

11. L. Luca Cavalli-Sforza, Paolo Menozzi, and Alberto Piazza, *The History and Geography of Human Genes* (Princeton, NJ: Princeton University Press, 1994); Michael F. Hammer, "A Recent Common Ancestry for Human Y Chromosomes," *Nature* 378, no. 6555 (1995): 376–78; Jobling and Tyler-Smith, "Fathers and Sons: The Y Chromosome and Human Evolution," 449–55; Karl Skorecki et al., "Y Chromosomes of Jewish Priests," *Nature* 385, no. 32 (1997): 32.

12. Jennifer K. Wagner and Kenneth M. Weiss, "Attitudes on DNA Ancestry Tests," *Human Genetics* 131 (2012): 41.

13. Charmaine D. Royal et al., "Inferring Genetic Ancestry: Opportunities, Challenges, and Implications," *American Journal of Human Genetics* 86 (2010): 661–73. See also Jennifer K. Wagner et al., "Tilting at Windmills No Longer: A Data-Driven Discussion of DTC DNA Ancestry Tests," *Genetics in Medicine* 14 (2012): 586. The DTC genetic-testing-services field is largely unregulated as an industry and for the most part privately owned. As a consequence, the companies are not obligated to be transparent about the number of customers they serve, the genetic markers or statistical algorithms they use, or their profits. What is known about this industry sector relies mostly on estimates and self-reporting.

14. For an incisive treatment of the "choices" and cognitive schema involved in genealogy, see Eviatar Zerubavel, *Ancestors and Relatives: Genealogy, Identity, & Community* (New York: Oxford University Press, 2012).

15. Strausbaugh et al., "The Genomics Perspective on Venture Smith," 227.

16. For an excellent, detailed discussion of this form of ancestry analysis, see Duana Fullwiley, "The Biologistical Construction of Race: 'Admixture' Technology and the New Genetic Medicine," *Social Studies of Science* 38 (2008): 695–735.

17. Wagner and Weiss, "Attitudes on DNA Ancestry Tests," 41.

18. David Lazer, introduction to *DNA and the Criminal Justice System: The Technology of Justice* (Cambridge, MA: MIT Press), 6.

19. Bolnick et al., "The Business and Science of Ancestry Testing."

20. Dahleen Glanton, "Blacks Tap Roots Anew," *Chicago Tribune*, September 12, 2004, http://articles.chicagotribune.com/2004-09-12/news/0409120142_1_reparations-movement-african-americans-slave-trade.

21. Ibid.

22. Darryl Fears, "Out of Africa—but From Which Tribe?," *Washington Post*, October 19, 2006, http://www.washingtonpost.com/wp-dyn/content/article/2006/10/18/AR2006101801635.html.

23. Strausbaugh et al., "The Genomics Perspective on Venture Smith," 215.

24. Paul Rabinow, *Essays on the Anthropology of Reason* (Princeton, NJ: Princeton University Press, 1996).

25. Harriott's autosomal DNA test inferred her racial composite results to be 28 percent European and 72 percent sub-Saharan African.

26. Martin Richards, "Beware the Gene Genies," *Guardian*, February 14, 2003, http://www.guardian.co.uk/comment/story/0,,899835,00.html.

27. Ibid.

28. Nash, "Genetic Kinship," 2004.

29. David M. Schneider, *American Kinship: A Cultural Account* (Englewood Cliffs, NJ: Prentice-Hall, 1968).

CHAPTER 5: ROOTS REVELATIONS

1. Parts of this chapter draw on collaborative research with Jeon Wong Hwang. See Nelson and Hwang, "Roots and Revelation: Genetic Ancestry Testing and the YouTube Generation," in *Race After the Internet*, ed. Lisa Nakamura and Peter Chow-White (New York: Routledge, 2013), 271–90.

2. The *Motherland* series also aired on cable television in the United States on the Sundance Channel.

3. June Deery, "Interior Design: Commodifying Self and Place in *Extreme Makeover*, *Extreme Makeover: Home Edition*, and *The Swan*," in *The Great American Makeover: Television, History, Nation*, ed. Dana Heller (New York: Macmillan, 2006), 169. On the history and affective significance of "the reveal" in television, see Anna McCarthy, "'Stanley Milgram, Allen Funt, and Me': Postwar Social Science and the 'First Wave' of Reality TV," in *Reality TV: Remaking Television Culture*, ed. Susan Murray and Laurie Ouellette (New York: New York University Press, 2004), 19–39.

4. Lawrence-Lightfoot and Jemison appear in *African American Lives*, produced by Henry Louis Gates Jr., Williams R. Grant, and Peter W. Kunhardt (Washington, DC: Public Broadcasting Service, 2005), DVD. Chris Rock's family genealogy is featured in *African American Lives II*, produced by Henry Louis Gates Jr., Williams R. Grant, Peter W. Kunhardt, and Dyllan McGee (Washington, DC: Public Broadcasting Service, 2007), DVD.

5. Isaiah Washington with Lavaille Lavette, *A Man from Another Land: How Finding My Roots Changed My Life* (New York: Center Street, 2011), 71.

6. Leslie Gordon, "Q & A with Isaiah Washington, a Man from Another Land," http://lesliewrites.com/isaiah.htm.

7. Isaiah Washington, "DNA Has Memory: We Are Who We Were," *Huffington Post*, http://www.huffingtonpost.com/isaiah-washington/dna-has-memory-we-are-who_b_87450.html.

8. Isaiah Washington, interview with author, March 2008.

9. Viewer responses described in this passage are posted in comments section, Jasmyne Cannick, "My African Ancestry DNA Revealed!," YouTube, http://www.youtube.com/watch?v=hZzQU3dT9DA&feature=related.

10. Ibid.

11. Ibid.

12. Edward Ball, *Slaves in the Family* (New York: Farrar, Straus and Giroux, 1998).

13. Leslie Goffe, *Priscilla: The Story of an African Slave*, BBC.com, November 23, 2005, http://news.bbc.co.uk/2/hi/africa/4460964.stm.

14. "Mission Statement," Gondobay Manga Foundation, https://gondobaymangafoundation.org/about-us/.

CHAPTER 6: ACTS OF REPARATION

1. Martha Biondi, "The Rise of the Reparations Movement," *Radical History Review* 87 (Fall 2003): 5.

2. Foner, *A Short History of Reconstruction* (New York: Harper & Row, 1990), 32.

3. Ibid., 72.

4. Ibid., 73.

5. Biondi argues that two factors account for the "growing" appeal of the reparations movement: (1) the model of successful settlements for past wrongs to victimized groups in the United States and abroad and (2) the sense that racial healing in the United States cannot be achieved "until the United States confronts the full scope of harms it inflicted on enslaved Africans and their descendants." Biondi, "The Rise of the Reparations Movement," 9.

6. Mary Frances Berry, "Reparations for Freedmen, 1890–1916: Fraudulent Practices or Justice Deferred?," *Journal of Negro History* 57 (July 1972): 220–21.

7. Johnson v. McAdoo, Opinion of the Court, November 14, 1916, *Cases Adjudged by the United States Court of Appeals (District of Columbia Circuit)*, vol. 45, pp. 440–41.

8. Ex-slave pension activists not only faced a tide of incredulity and recalcitrance from elected officials, but also repressive state surveillance. As occurred simultaneously with race leader Marcus Garvey and other black activists of the period, the leadership of the ex-slave association was subject to scrutiny and harassment from federal authorities. In pathbreaking research, historian Mary Frances Berry documented how House and other leaders of the movement were falsely accused of being engaged in fraud by taking money gathered from the organization membership and using these monies for their personal ends. These accusations were unsupported by evidence but would result in a trial and conviction for House on fraud charges that precipitated the end of the organization.

9. Cheryl Townsend Gilkes, "Interview with Queen Mother Audley

Moore," in *The Black Women Oral History Project*, vol. 8, ed. Ruth Edmonds Hill (Westport, CT: Meckler, 1991), 115.

10. Lelya Keough, "Moore, 'Queen Mother' Audley," in *Africana: The Encyclopedia of African and African American Experience*, ed. Kwame Anthony Appiah and Henry Louis Gates Jr. (New York: Oxford University Press, 2005), 59–60.

11. United Nations, Convention on the Prevention and Punishment of the Crime of Genocide, http://www.hrweb.org/legal/genocide.html.

12. Townsend Gilkes, "Interview with Queen Mother Audley Moore," 115.

13. Queen Mother Moore, http://www.queenmothermoore.org/reparations.htm, accessed April 13, 2009; Erik S. McDuffie, "Moore, Audley 'Queen Mother,'" *American National Biography Online*, August 2003, http://www.anb.org/articles/15/15–01298.html.

14. Biondi, "The Rise of the Reparations Movement," 7.

15. Ibid.

16. Julie Foster, "Slavery Reparations Lawsuit Brewing," January 31, 2001, *WND*, http://www.wnd.com/2001/01/8011/#7KkKpIXfEZV8jsTF.99.

17. The RNA sought to create a new world for African slave descendants while appealing to another facet of international law—the right to autonomous rule. The RNA demanded reparations from the United States in the form of $400 billion in "slavery damages." In tandem, these activists, like the Black Panther Party before them, invoked UN statutes to call for the formation of a "plebiscite" among black Americans. The purpose of this poll would be to decide whether members of the community would establish an independent nation with territory ceded from the United States. They wanted African Americans to have the choice of citizenship beyond the United States, and accordingly proposed the creation of a separate nation for blacks carved out of several southern states, comprising in part some of the land promised to emancipated men and women after the Civil War. The RNA thereby added secession to the diverse arsenal of reparations strategies.

18. The text of this proclamation was subsequently printed in the *New York Review of Books*. See the Black National Economic Conference, "Black Manifesto," *New York Review of Books*, July 10, 1969, http://www.nybooks.com/articles/archives/1969/jul/10/black-manifesto/.

19. Ibid.

20. John Torpey, "Paying for the Past? The Movement for Reparations for African Americans," *Journal of Human Rights* 3 (2004): 173.

21. Biondi, "The Rise of the Reparations Movement," 9.

22. Ta-Nehisi Coates, "The Case for Reparations," *Atlantic*, May 2014, http://www.theatlantic.com/features/archive/2014/05/the-case-for-reparations/361631/.

23. Despite the national evasion of this foundational issue, recent public

acknowledgment of and apologies for slavery have occurred in several states and municipalities, including California, Maryland, and Illinois. In 2000, California governor Gray Davis signed a law requiring insurance companies operating in the state to disclose any policies that they wrote on enslaved men and women. In 2002, the California Department of Insurance released a report revealing that six insurance companies now operating in the state had profited from slavery. The case for reparations was also taken up by several state and city governments, including Chicago, where, since 2003, companies seeking to do business with the city are required to disclose whether they profited from chattel slavery.

24. Emphasis added. United States Court of Appeals, Ninth Circuit, Jewel Cato; Joyce Cato; Howard Cato; Edward Cato, Plaintiffs-Appellants, v. United States of America, Defendant-Appellee. Leerma Patterson; Charles Patterson; Bobbie Trice Johnson, et al., Plaintiffs-Appellants, v. United States of America, Defendant-Appellee. Nos. 94–17102, 94–17104. Decided December 4, 1995.

25. Charles J. Ogletree Jr., "Tulsa Reparations: The Survivors' Story," http://www.bc.edu/content/dam/files/schools/law/lawreviews/journals/bctwj/24_1/03_TXT.htm.

26. Randall Robinson, *The Debt: What America Owes to Blacks* (New York: Dutton, 2000), 243.

27. Ibid.

28. Tamar Lewin, "Calls for Slavery Restitution Getting Louder," *New York Times*, June 4, 2001, http://www.nytimes.com/2001/06/04/national/04SLAV.html.

CHAPTER 7: THE ROSA PARKS OF THE REPARATIONS LITIGATION MOVEMENT

1. Raymond A. Winbush, ed., *Should America Pay? Slavery and the Raging Debate on Reparations* (New York: Amistad, 2003), 31; "Personal Testimony of Deadria Farmer-Paellmann in Support of HR 40," Washington, DC, April 6, 2005, http://ncobra.org/resources/pdf/DEADRIA%20april%206%20statement.pdf.

2. Deadria C. Farmer-Paellmann, "Excerpt from *Black Exodus: The Ex-Slave Pension Movement Reader*," in Winbush, *Should America Pay?*, 22.

3. Farmer-Paellmann, "Excerpt from *Black Exodus*," 24.

4. Robert Trigaux, "Putting a Price on Corporate America's Sins of Slavery," *St. Petersburg Times*, April 14, 2002.

5. Ibid., Farmer-Paellmann, "Excerpt from *Black Exodus*," 26.

6. Farmer-Paellmann, "Excerpt from *Black Exodus*," 24.

7. Ibid.

8. Robin Finn, "Public Lives: Pressing the Cause of the Forgotten Slaves," *New York Times*, August 8, 2000.

9. "Personal Testimony of Deadria Farmer-Paellmann in Support of HR 40."

See also James Cox, "Farmer-Paellmann Not Afraid of Huge Corporations," *USA Today*, February 21, 2002, http://www.usatoday.com/money/general/2002/02/21/slave-activist.htm.

10. Farmer-Paellmann, "Excerpt from *Black Exodus*," 25. See also "Personal Testimony of Deadria Farmer-Paellmann in Support of HR 40," and Cox, "Farmer-Paellmann Not Afraid of Huge Corporations."

11. Trigaux, "Putting a Price on Corporate America's Sins of Slavery."

12. "Personal Testimony of Deadria Farmer-Paellmann in Support of HR 40."

13. Ibid.

14. Farmer-Paellmann, "Excerpt from *Black Exodus*," 26.

15. Trigaux, "Putting a Price on Corporate America's Sins of Slavery."

16. "Personal Testimony of Deadria Farmer-Paellmann in Support of HR 40."

17. Josie Appleton, "Suing for Slavery," April 1, 2004, Spiked-Liberties, http://www.spiked-online.com/Printable/0000000CA4B0.htm.

18. Trigaux, "Putting a Price on Corporate America's Sins of Slavery."

19. Norman Kempster, "Agreement Reached on Nazi Slave Reparation," *Los Angeles Times*, December 15, 1999, http://articles.latimes.com/1999/dec/15/news/mn-44055.

20. Biondi, "The Rise of the Reparations Movement," 5.

21. Ibid., 9.

22. For additional discussion of the Farmer-Paellmann case, see Faubion and Hamilton, "Sumptuary Kinship."

23. Cox, "Farmer-Paellmann Not Afraid of Huge Corporations."

24. Ibid.

25. Peter Viles, "Suit Seeks Billions in Slave Reparations," CNN.com, March 29, 2002; Matt O'Connor, "Judge Drops Suit Seeking Reparations, Slave Descendants Vow to Appeal," *Chicago Tribune*, January 27, 2004.

26. In 2008 Fagan was disbarred in the states of New York and New Jersey for embezzling and misappropriating funds owed to clients in the Nazi reparations case.

27. Quoted in Robert Trigaux, "Putting a Price on Corporate America's Sins of Slavery."

28. Ibid. See also Nathan Burchfiel, "Blacks Deserve '200 Years of Free Education,' Activist Says," TownHall.com, http://www.townhall.com/news/ext_wire.html?rowid=46239.

29. Kelly Vlahos Beaucar, "Lawsuit Chases Companies Tied to Slavery," FOXNews.com, March 27, 2002, http//www.foxnews.com/printer_friendly_story/0,3566,48781,00.html.

30. Finn, "Public Lives."

31. Faubion and Hamilton, "Sumptuary Kinship," 553.

32. Emphasis added, In Re African-American Slave Descendants Lit., 304 F. Supp. 2d 1027 (N.D. Ill. 2004), Opinion and order.

33. Ibid. See also O'Connor, "Judge Drops Suit Seeking Reparations."

34. Eric J. Miller, "Representing the Race: Standing to Sue in Reparations Lawsuits," *Harvard BlackLetter Law Journal* 20 (2004): 93, http://www.law.harvard.edu/students/orgs/blj/vol20/ericmiller.pdf.

35. Alex Kleiderman, "The Vexed Question of Paying for Slavery," BBC News, August 23, 2004, http://news.bbc.co.uk/go/pr/fr/-/1/hi/business/3590334.stm.

36. "Students to Boycott Slavery Banks; Marks 50th Anniversary of Montgomery Bus Boycott," press release, Restitution Study Group, December 5, 2005, http://releases.usnewswire.com/printing.asp?id=57572.

37. Burchfiel, "Blacks Deserve '200 Years of Free Education.'"

38. "Do Not Use Aetna as Your Health Insurance Carrier," e-mail from Deadria Farmer-Paellmann to OTUAfricanDNA Yahoo Group, November 7, 2006.

39. Appleton, "Suing for Slavery."

40. Ibid.

41. "Business in Brief: Slaves' Descendants File $1B Lawsuit," *Boston Globe*, March 30, 2004.

42. Farmer-Paellmann v. FleetBoston Refiling: African American Genocide and DNA Case, 16.

43. James Cox, "Lloyd's of London, FleetBoston and R.J. Reynolds Accused; Descendants Seek $1 Billion," *Miami Times*, March 31–April 6, 2004.

44. Nick Godt, "J.P. Morgan & Co. Sued for Profiting from Slavery," *New York Sun*, September 26, 2006, http://www.nysun.com/article/40357. See also Faubion and Hamilton, "Sumptuary Kinship," 533.

45. Biondi, "The Rise of the Reparations Movement," 14.

46. Kevin Hopkins, "Forgive US Our Debts? Righting the Wrongs of Slavery," *Georgetown Law Journal* (August 2001).

47. In Re African-American Slave Descendants Litigation 304 F. Supp. 2d 1027 (N.D. Ill. 2005), Opinion and order.

48. Ashley M. Heher, "Slave Descendants Attempt to Revive Reparations Suit," *Chicago Sun Times*, September 27, 2006.

49. Faubion and Hamilton, "Sumptuary Kinship," 552.

50. Orlando Patterson, *Slavery and Social Death: A Comparative Study* (Cambridge, MA: Harvard University Press, 1985).

51. Deadria Farmer-Paellmann interview with the Australian Broadcasting Corporation, "The World Today," March 30, 2004.

52. Adolph Reed, "The Case Against Reparation," *Progressive*, December 2000.

53. Ibid.

54. Winston Munford, "How Do We Make Reparations Happen?," *Black*

Star News, January 28, 2014, http://www.blackstarnews.com/us-olitics/justice/how-do-we-make-reparations-happen.html.

55. O'Connor, "Judge Drops Suit Seeking Reparations."

56. Kelly Vlahos Beaucar, "Lawsuit Chases Companies Tied to Slavery," FOXNews.com, March 27, 2002, http//www.foxnews.com/printer_friendly_story/0,3566,48781,00.html.

57. James Davey, "From 'Jim Crow' to 'John Doe': Reparations, Corporate Liability, and the Limits of Private Law," in *Ethics, Law, and Society*, vol. 3, ed. Jennifer Gunning and Soren Holm (Burlington, VT: Ashgate, 2007), 199.

58. Glanton, "Blacks Tap Roots Anew."

59. Ibid.

CHAPTER 8: DNA DIASPORAS

1. Rogers Brubaker, "The 'Diaspora' Diaspora," *Ethnic and Racial Studies* 28, no. 1 (2005): 1.

2. On "exceptional," see Khachig Tölölyan, "Rethinking Diaspora(s): Stateless Power in the Transitional Moment," *Diaspora* 5 (1996): 13. Also see Gomez, *Exchanging Our Country Marks*; Robin D. G. Kelley and Tiffany Patterson (2000). "Unfinished Migrations: Reflections on the African Diaspora and the Making of the Modern World," *African Studies Review*, 43, no. 1 (2000): 11–45.

3. William Safran, "Diasporas in Modern Societies: Myths of Homeland and Return," *Diaspora* 1 (1990): 91.

4. Janet Carsten, *Cultures of Relatedness: New Approaches to the Study of Kinship* (Cambridge, UK: Cambridge University Press, 2000).

5. Carol Stack, *All Our Kin: Strategies for Survival in a Black Community* (New York: Harper & Row, 1974).

6. Laurence J. C. Ma, "Space, Place, and Transnationalism in the Chinese Diaspora," in *The Chinese Diaspora: Space, Place, Mobility, and Identity*, ed. Laurence J. C. Ma and Carolyn Cartier (Lanham, MD: Rowman & Littlefield, 2003), 1–50.

7. Jacqueline Nassy Brown, "Black Liverpool, Black America, and the Gendering of Diasporic Space," *Cultural Anthropology* 13 (1998): 298; Jacqueline Nassy Brown, *Dropping Anchor, Setting Sail: Geographies of Race in Black Liverpool* (Princeton, NJ: Princeton University Press, 2005), 53.

8. Brown, "Black Liverpool, Black America," 298.

9. Rukmini Callimachi, "West African Nation Lays Claim to Whoopi," Associated Press, February 7, 2007.

10. "US Chat Show Host Could Be a Zulu," BBC News, June 15, 2005, http://news.bbc.co.uk/2/hi/africa/4096706.stm.

11. "Liberia: Billionaire Showcases Liberia," AllAfrica.com, May 1, 2006, http://allafrica.com/stories/200605010729.html.

12. "The Leon H. Sullivan Foundation," Leon H. Sullivan Foundation, http://www.thesullivanfoundation.org/summit/about/index.asp.

13. "About the Summit," Leon H. Sullivan Foundation, http://www.thesullivanfoundation.org/summit/about/index.asp.

14. Leon H. Sullivan, *Build, Brother, Build* (Philadelphia: Macrae Smith, 1969), 52.

15. Zion Baptist Church of Philadelphia, http://www.zionbaptphilly.org/index.php/about-us/about-us-who-we-are.

16. Leon H. Sullivan, *Moving Mountains: The Principles and Purposes of Leon Sullivan* (Valley Forge, PA: Judson Press, 1998).

17. Ibid., 27–30.

18. Leon H. Sullivan, "Agents for Change: The Mobilization of Multinational Companies in South Africa," *Law & Policy in International Business* 15 (1983): 429. Karen Paul, "Corporate Social Monitoring in South Africa: A Decade of Achievement, An Uncertain Future," *Journal of Business Ethics* 8 (1989): 465.

19. Paul Lewis, "Leon Sullivan, 78, Dies; Fought Apartheid," obituary, *New York Times*, April 26, 2001, http://www.nytimes.com/2001/04/26/world/leon-sullivan-78-dies-fought-apartheid.html,

20. Paul, "Corporate Social Monitoring in South Africa," 464.

21. Lewis, "Leon Sullivan, 78, Dies."

22. Sullivan, *Moving Mountains*, 28 (emphasis added).

23. Leon H. Sullivan Foundation, "Framework for the Establishment of Africa-Diaspora Dual Citizenship" (Washington, DC: Sullivan Foundation, n.d.).

24. Gregory Simpkins, interview with author, July 3, 2012.

25. Ibid.

26. Ibid.

27. Leon Sullivan Foundation, "Framework for the Establishment of Africa-Diaspora Dual Citizenship."

28. Paige, interview with author.

29. Simpkins, interview with author.

30. Ibid.

31. Ibid.

32. Paige, interview with author.

33. Ibid.

34. Chad Bouchard, "Leon H. Sullivan Foundation: The Implosion of a Legacy," *Washington Post Magazine*, July 25, 2013, http://www.washingtonpost.com/lifestyle/magazine/leon-h-sullivan-foundation-the-implosion-of-a-legacy/2013/07/18/fe042654-d9ba-11e2-a9f2-42ee3912ae0e_story.html.

35. Ibid.

CHAPTER 9: RACIAL POLITICS AFTER THE GENOME

1. Gregory Simpkins interview with author, July 3, 2012. "They had already taken the test. . . . Gina knew that, and we were brainstorming about people [who] had already taken this . . . [and] so, they're already open to it. And [Garvey and King III] are obviously a draw. That's how it came about."

2. Alvin M. Weinberg, "Science and Trans-Science," in his *Nuclear Reactions: Science and Trans-Science* (New York: American Institute of Physics, 1992), 3–20.

3. Mapping Police Violence, "The National Police Violence Map," www.mappingpoliceviolence.org.

INDEX